Bioinformatics

# Also of interest

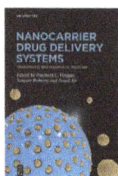

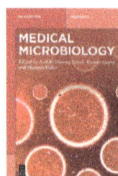

# Bioinformatics

Drug Discovery

Edited by
Anil K. Sharma and Varruchi Sharma

**DE GRUYTER**

**Editors**
**Anil Kumar Sharma**
Department of Biotechnology
Amity University
Sector 82 A
IT City Rd
Block D
Mohali 140306
Punjab
India
anibiotech18@gmail.com

**Varruchi Sharma**
Department of Biotechnology and Bioinformatics
Sri Guru Gobind Singh College
Sector 26
Chandigarh 160019
India
sharma.varruchi@gmail.com

ISBN 978-3-11-156788-4
e-ISBN (PDF) 978-3-11-156858-4
e-ISBN (EPUB) 978-3-11-156882-9

**Library of Congress Control Number: 2024944101**

**Bibliographic information published by the Deutsche Nationalbibliothek**
The Deutsche Nationalbibliothek lists this publication in the Deutsche Nationalbibliografie;
detailed bibliographic data are available on the internet at http://dnb.dnb.de.

© 2025 Walter de Gruyter GmbH, Berlin/Boston
Cover image: metamorworks/iStock/Getty Images Plus
Typesetting: Integra Software Services Pvt. Ltd.

www.degruyter.com

# Contents

# Contributors

**Nitin Sharma**
Department of Biotechnology
Chandigarh Group of Colleges
Landran, Mohali
Punjab
India

**Manika Choudhary**
Department of Biotechnology
Chandigarh Group of Colleges
Landran, Mohali
Punjab
India

**Vikas Kumar**
Faculty of Applied Sciences and Biotechnology
Shoolini University
Solan
Himachal Pradesh
India

**Varruchi Sharma**
Department of Biotechnology & Bioinformatics
Sri Guru Gobind Singh College
Sector 26
Chandigarh 160019
India
sharma.varruchi@gmail.com

**Imran sheikh**
Department of Biotechnology
Eternal University
Baru Sahib
Sirmour
Himachal Pradesh
India

**Vikas Kushwaha**
Department of Biotechnology and Bioinformatics
Sri Guru Gobind Singh College
Sector 26
Chandigarh 160019
India

**Shagun Gupta**
Department of Bio-Sciences and Technology
MMEC
Maharishi Markandeshwar (Deemed to be
University)
Mullana, Ambala 133207
Haryana
India

**Ankur Kaushal**
Department of Bio-Sciences and Technology
MMEC
Maharishi Markandeshwar (Deemed to be
University)
Mullana, Ambala 133207
Haryana
India

**Seema Ramniwas**
University Centre for Research and Development
University Institute of Biotechnology
Chandigarh University
Gharuan, Mohali
Punjab
India

**Poonam Bansal**
Department of Bio-Sciences and Technology
MMEC
Maharishi Markandeshwar (Deemed to be
University)
Mullana, Ambala 133207
Haryana
India

**Anupam Sharma**
Department of Physics
Guru Kashi University
Talwandi Sabo, Bathinda
Punjab
India

https://doi.org/10.1515/9783111568584-203

**Vandana Sharma**
Department of Physics
MMEC
Maharishi Markandeshwar (Deemed to be University)
Mullana, Ambala 133207
Haryana
India

**J. K. Sharma**
Department of Physics
MMEC
Maharishi Markandeshwar (Deemed to be University)
Mullana, Ambala 133207
Haryana
India

**Anil Panwar**
Department of Bioinformatics and Computational Biology
College of Biotechnology
CCS Haryana Agricultural University
Hisar
Haryana
India

**Anil Kumar Sharma**
Department of Biotechnology
Amity University
Sector 82 A
IT City Rd, Block D
Mohali 140306
Punjab
India
anibiotech18@gmail.com

**Damanjeet Kaur**
Department of Biotechnology and Bioinformatics
Sri Guru Gobind Singh College
Sector 26
Chandigarh 160019
India

**Amit Joshi**
Department of Biotechnology and Bioinformatics
Sri Guru Gobind Singh College
Sector 26
Chandigarh 160019
India

**Sonal Datta**
Department of Bio-Sciences and Technology
Maharishi Markandeshwar (Deemed to be University)
Mullana, Ambala 133207
Haryana
India

**Ramesh C. Thakur**
Department of Chemistry
Himachal Pradesh University
Summer Hill
Shimla 171005
Himachal Pradesh
India
drthakurchem@gmail.com

**Akshay Sharma**
Department of Chemistry
Himachal Pradesh University
Summer Hill
Shimla 171005
Himachal Pradesh
India

**Renuka Sharma**
Department of Chemistry
Himachal Pradesh University
Summer Hill
Shimla 171005
Himachal Pradesh
India

**Meenakshi Rajpoot**
Department of Biotechnology
DAV College (Lahore)
Ambala City
Ambala
Haryana
India

**Anu Prabha**
Department of Botany
GGDSD College Sector 32
Chandigarh

**Sheetal Dagar**
Department of Bioinformatics and
Computational Biology
CCS Haryana Agricultural University
Hisar 125004
Haryana
India

**Sri Kant**
Academic Affairs Department
Chandigarh University
Gharuan
Mohali 140413
Punjab
India

Nitin Sharma, Manika Choudhary, and Vikas Kumar

# Chapter 1
# Introduction to bioinformatics

Bioinformatics has been able to establish and bridge the gap between modern biology and other disciplines through computational means. With the vast amount of biological data on hand, its precise quantification and analysis are required. Complete and reliable annotation of database sequence information with accurate biological data has been made possible through the advent of bioinformatics. Through this field, errors in DNA and protein sequences can be understood more precisely. Predictive information can be derived to design experiments to quickly and reliably verify the biology behind the sequence. The main idea of bioinformatics is the integration of biology into the computational field by collecting the data in one place, where it is easily accessible to all users through online access by websites that organize data. Despite advancements in this field, there is still a need for effective tools to perform better annotation with high accuracy and improve the quality of sequenced genomes without gaps. Moreover, advanced bioinformatics tools need to be developed for the analysis of growing high-throughput metagenomics, proteomics, and metabolomics.

## 1.1 Introduction

Bioinformatics has become a major driving force in research in the field of biological sciences, bridging the gap between modern biology and other disciplines. Modern biology needs quantification and analysis, which is why bioinformatics plays a significant role in comprehensive studies of the storage of DNA data, mathematical and statistical modeling, and computational molecular biology. However, bioinformatics alone involves the perusal of the sequence, structure, and function of genes, whereas the biological processes are facilitated by tools or algorithms in computation.

Bioinformatics is widely encompassing through the majority of scientific fields nowadays, making this branch difficult to define precisely due to its wide-ranging applications. However, it can be understood with the use of computers and their tools to analyze the biological macromolecules' information, store data, distribute information, and retrieve them later. Bioinformatics helps in determining the structural, functional, and phylogenetic relationships between biological macromolecules such as

**Nitin Sharma, Manika Choudhary**, Department of Biotechnology, Chandigarh Group of Colleges, Landran, Mohali, Punjab, India
**Vikas Kumar**, Faculty of Applied Sciences and Biotechnology, Shoolini University, Solan, Himachal Pradesh, India, e-mail: abhinitu30@gmail.com

https://doi.org/10.1515/9783111568584-001

DNA, RNA, and proteins. According to the National Center for Biotechnology Information, bioinformatics is defined as the application of computational and analytical tools to capture and interpret biological data. It is an interdisciplinary field that harnesses fields comprising computer science, mathematics, biology, and physics.

The term "bioinformatics" was coined by Paulien Hogeweg and Ben Hesper in 1970 and was widely defined as the informatics processes in the biotic system. From the late 1980s onward, the term "bioinformatics" has mostly been used to refer to the comparative analysis of genome data by computational methods. This dates back to the 1860s when the evolution from a gene to today's genome analysis has been achieved through this very important tool and technology. The concept of passing a material from the parent to offspring became a wider area of attention when Gregor Mendel discovered the principles of genetic inheritance. During that period until the 1950s, there was no appropriate information about DNA, its actual location, and its recognition as a carrier molecule was also contentious.

Margaret Dayhoff (1925–1983) was the first bioinformatician who is also known as the founder of bioinformatics. She introduced computational techniques in the sequencing of proteins and nucleic acids and established the first publicly available database for research, known as "Comprotein" which computes the primary structure of proteins.

The integration of bioinformatics and computational biology encompasses the entire collection of data optimization and management systems through biological databases and tools, which analyze large sets of data and help in sequencing the human genome, further assisting in extracting out information from proteomics and genomics.

## 1.2 Biological databases

Biological databases are software programs to which biological data is submitted, including sequence information, protein structural data, graphs indicating relationships, genome patterns, and additional macromolecule-related information. The main idea of bioinformatics is integration by collecting the data in one place where it is easily accessible to all users through online access via websites that organize data. The first database to arise was GenBank, which is a collection of all available proteins and DNA sequences. Subsequently, an enormous number of databases were created, including NCBI, DDBJ, and EMBL, according to the operation to be executed and based on the type of biological macromolecule. At present, a total of 1,552 databases are publicly accessible online, according to a 2014 report published in the journal *Nucleic Acids Research*. Summarizing the above, and more precisely, databases are programs that are (i) used for indexing, organizing data, and maintaining it in a constant state for infinite time and (ii) accessible to all researchers and users to find relevant data for their work in one place, saving time, work, and resources.

Availability of the data is another crucial element for an easy approach to biological data, which is accessible in mainly three forms: text format, sequence data, and protein structures. Text formats are given in PubMed, which is basically an online access to the MEDLINE databases maintained by the National Library of Medicine. It has free online journals and databases to retrieve information by using keywords. Multiple keywords, author names, and journal titles are other search criteria that are used for easy retrieval of information. Another example is OMIM, an abbreviation for Online Mendelian Inheritance in Man, which is consistently used by medical investigators for the interpretation of genetic disorders. It is an online catalog of all known human genes, genetic disorders, and genetic phenotypes.

Sequence data are typified by GenBank, which is produced and maintained by the National Center for Biotechnology Information. It provides an online access to nucleotide sequences and their protein translations with appropriate biological annotation. Only original sequences can be submitted to GenBank, as it examines the originality of the data first and then assigns it an accession number. UniProt stands for "universal protein resource," which is an online public database of all protein sequences, consisting of a vast amount of information about the biological function of proteins. The UniProt association is a collaboration between the European Bioinformatics Institute (EBI), the Protein Information Resource, and the Swiss Institute of Bioinformatics. It serves as the central repository of protein sequences at EBI.

Protein structures are provided by SCOP and CATH, which describe the structural and evolutionary relationships between all known protein structures and classify all protein structures derived from PDB, respectively. Huge databases are broadly classified into primary, secondary, and composite databases.

**Primary databases** are also known as archival databases. They contain raw data of the structure or sequence of nucleic acids and proteins, which are experimentally derived by researchers. The submitted data is essentially archival in nature. An accession number is given to each and every data submission, after which they form part of the scientific record. Examples are GenBank and DDBJ (nucleotide sequence), PDB (three-dimensional structures), and EMBL (genome and nucleotide sequence).

**Secondary databases** comprise data derived from the computational postprocessing of raw primary data. They contain additional annotations and biological information regarding the functions and structures of macromolecules. They often provide additional information to the protein sequence derived from their own analysis, featuring its particular characteristics. They store information such as conserved sequences, sites for protein modification, or active site residues. UniProt (sequence and functional information on proteins) and Ensembl (variation, function, and regulation of whole genome sequences) are two such examples of secondary databases.

**Composite databases** are a combination of various primary database resources. This helps in lessening the tedious task of searching through multiple databases referring to the same data. The approach used, for instance, the search algorithm employed, differs considerably in every composite database. For example, DrugBank offers details on drugs and their targets, BioGraph incorporates assorted knowledge of biomedical science, and BioModel is a storehouse of computational models of biological developments. There are many composite databases that provide users with various tools and software for data analysis. NCBI, being a composite database, has stored sequences of nucleotides and proteins within its server and thereby suffers from high redundancy in the data deposited (bioinformatics data resources: https://www.researchgate.net/publication/31382276).

## 1.3 Bioinformatics tools

Bioinformatics tools are search programs designed to identify the classification and potential homologs from a particular sequence after it has been saved and analyzed. These tools aid in extracting meaningful results from the saved raw biological data. These tools are operated for the following results:
- Sequence alignment
- Database similarity
- Multiple sequence alignment
- Prediction of protein motifs and domains
- Molecular phylogenetics

The prominent sequence similarity search tools are BLAST and FASTA available on the web. By using these tools, potential homologs can be identified, which are then used for a better understanding of the query sequences and in the prediction of structures or aiding in phylogenetic analysis. Global and local alignments are the two alignment strategies for identifying the regions of similarity among residues between two sequences. Global alignment refers to the best possible alignment across the entire length, whereas in local alignment, only the local region with the highest level of similarity between residues aligns. Alignment algorithms include dot matrix, dynamic programming, and the word method.

Clearly, bioinformatics can now be literally defined as the application of computer technology or information that merges applied mathematics, statistics, computer science, and biology to understand and manage biological data for seeking hidden information in the large volume of information. For instance, in the genome project, over a thousand genomes have now been sequenced, which include viral, organellar, bacterial chromosomes, and plasmids. Due to the elevation in research in the field of biotechnology and molecular biology, an immense amount of nucleotide data is being produced. To organize, analyze, and manage this tremendous data requires efficient problem-solving power, which is why computers have a major role in assisting with biological information. This solves the major problems of today's era, such as human health, environmental issues, the agriculture sector, and energy resources.

Until now, it has been understood how bioinformatics plays a significant role in decoding the code of life, which is DNA and genomes. The application on the broad spectrum includes structure, function, and sequence analysis, which integrates into large field applications such as biotechnology, life sciences, biomedical sciences, agriculture, and the atmosphere.

## 1.4 Scope of bioinformatics

Field with its major expansion and revolutionization providing next-generation sequencing analysis certainly has taken a high leap in biological research and more precisely in molecular research. For instance, the efforts on whole genome sequencing by approaching third-generation sequencing have provided a huge information regarding the genes and their functions in human genomes, which has advanced the technology in genetic testing, drug design, and gene therapy. This will also ultimately help to improve orthologous gene identification tools that currently need attention.

It has become an essential interdisciplinary scientific platform assisting the "omics" field. The generation of high-throughput biological data due to the growth in the omics field has largely demanded the need for bioinformatics resources. The primary target is to understand the biological processes at the molecular level to interpret the knowledge from the treasure of the central dogma of life.

## 1.5 Applications of bioinformatics

### 1.5.1 Medicine

*Molecular medicine*: After the completion of the Human Genome Project, much has been learned about genes and their influence on human diseases. Genetic diseases may be inherited like cystic fibrosis or may be a consequence of the body's response to environmental stress, which causes alterations in the genome; for example, diabetes, cancers, and heart disease. As the complete human genome is drafted, the molecular basis of a disease can be found by searching for the genes directly related to it. It is much easier to find cures and better treatments with the knowledge of the molecular mechanisms of diseases.

*Gene therapy*: Genes themselves are used to treat diseases by introducing a functional gene into the cells of the patient in order to correct the disease. Gene therapy is used to cure, treat, or prevent a certain disease by changing the gene expression.

*Predictive medicine*: Preventive measures or processes are conducted by expecting the probability of a disease, and they either prevent the disease or decrease its effects on the patient, for example, prenatal testing and newborn screening.

*Personalized medicine*: The development of targeted therapeutic drugs and accompanying diagnostics serves the purpose of successful therapy. The size of patient groups that a therapy is projected to be successful for, notwithstanding the paucity of such trials. By analyzing a person's genetic profile, the best drug therapy can be used to cure or prevent the disease, as the relationship between the disease phenotype and the molecular data of a patient is specific, so a generalized drug does not always show the desired results. Thus, bioinformatics plays a major role in understanding the molecular data and helping the practicing physician in opting for the best therapy available in the market. HIV and cancer are the two substantial fatal diseases for which bioinformatics-assisted therapy is being used.

Drug discovery and the development of drugs specific to the target aid in more successful treatments and fewer side effects. Bioinformatics provides strategies and algorithms to validate new drug targets and to store and manage available drug target information.

## 1.5.2 Microbial genome applications

*Waste cleanup*: Bacteria and microbes play a major role in cleaning waste as there are microorganisms that remediate pollutants into the natural biogeochemical cycle. This approach is called bioremediation, which utilizes the microbial capability for biodegradation. This technology is further improved by the use of bioinformatics by analyzing the genomes of microbes and the structural characteristics of proteins. It helps in understanding the mechanisms of biodegradative pathways of microbes. One such example is *Deinococcus radiodurans*, a radiation-resistant organism used for the cleanup of toxic metals and radioactive waste sites.

*Biotechnology*: The use of bioinformatics in biotechnology mainly signifies the industrial field, where the knowledge of physiology and the genetic makeup of bacteria can be optimally utilized, especially in food and pharmaceutical industries.

Energy sources around the globe have been an issue of major concern. Bioinformatics has greatly helped to expound on this problem. Microbes like *Chlorobaculum tepidum*, a low-light-adapted photoautolithotrophic sulfur-oxidizing bacterium, have evolved to cope with changing energy availability by tuning their proteome for energy efficiency. By studying the genome and proteomics of the microbe, they can be used for generating energy from light.

## 1.5.3 Agriculture

*Insect resistance*: Plants and crops need to be resistant to yield high productivity with higher nutritional value. In order to achieve the desired traits, scientists mapped the genes of the bacterium *Bacillus thuringiensis* and used its genes to incorporate into the plant genome to make it resistant to insects. Thus, when pests eat the plants, the bacterium toxin enters into their bloodstream and kills them. This approach has helped agriculturists reduce the use of pesticides and enhance the quality of the crops.

With improved nutritional quality, golden rice is the genetically modified crop that is produced by the transfer of genes to elite crop varieties. Understanding the genomes of the plant and editing them using advanced molecular techniques facilitated precise, efficient, and targeted modifications at genomic loci, which could eradicate the problem of people suffering from malnourishment. By the insertion of genes into plant genomes, the nutritional value of plants like tomatoes could be increased. Similarly, inserting a gene derived from yeast into a tomato could produce fruits having an extended shelf life.

Stress-resistant plants and stress-tolerant varieties can be developed by identifying the stress tolerance genes and alleles. Various tools have been developed to study physiology, expression profiling, and comparative genomics. The KEGG database contains all the metabolic pathways, including the pathway for carbohydrate production. Genes in the ABA production pathway are important for the development of drought-resistant varieties. The KEGG database is important in identifying the pathways for carbohydrate and ABA production.

After the identification of the pathways, the genes involved in the same pathway are further studied. Progress has been made in developing cereal varieties that have a greater tolerance for soil alkalinity and are free of aluminum and iron toxicities. These varieties will allow agriculture to succeed in poorer soil areas, thus adding more land to the global production base. Research is also in progress to produce crop varieties capable of tolerating reduced water conditions.

## 1.5.4 Bioweapon creation

This includes the analysis of pathogen genomes and emerging methods involving knowledge management and their impact on intelligence. It is the intentional release of pathogens for medical intelligence, forensic operations, and biothreat awareness. Anthrax is one of the known instances of the most deadly agents to be used as a bioweapon. Scientists have also recently built the virus poliomyelitis using entirely artificial means.

### 1.5.5 Evolutionary studies

Evolutionary relationships among individuals or groups of organisms can be found out by comparing their genomes or by constructing a phylogenetic tree based on sequence alignments using algorithmic tools. Bioinformatics simplifies phylogenetics, whereas using anatomical methods to find evolutionary relationships is very time-consuming.

### 1.5.6 Forensic science

Bioinformatics-based tools are taking over the forensic sciences and research, whether it be DNA profiling, gender identification, extracting the tiniest DNA from the minimal microbial sample, or the most advanced automatic fingerprint recognition system. Moreover, next-generation sequencing technology and epigenetic studies give rise to a vast research domain by simultaneously analyzing multiple loci that are of forensic interest in multiple genetic contexts. Bioinformatics has a great impact on making hypotheses on DNA samples with the help of forensic statistics.

### 1.5.7 Antibiotic resistance

Antibiotics work as a selective pressure in the spread of resistance genes through the exchange of genetic material among bacterial isolates. Bioinformatics studies have provided ways of modeling biological living cell systems and proteins, enabling scientists to discover effective drug strategies.

## 1.6 Limitations

Biological data is highly complex. It has richness and enormity of information that is sometimes hard to maintain as the inclining researches in genetics and molecular biology are speeding up at a very high rate. To store, organize, and manage the immense amount of data require flexibility in handling and in developing newer, faster algorithm tools and software.

Most biologists will not care or know about the data structure or the scheme design. Thus, the interface to the biological database/resource should display information to the user in a manner appropriate for the problem being addressed and that reflects the underlying data structures. That is, database access should be through a transparent, intuitive user interface. Usually, the presentation of the same data differs among biologists, as data can be portrayed in many models. This causes heterogeneity in how data are analyzed, annotated, and displayed. The use of computers for information storage and retrieval of large amounts of data saves time, cost, and wet lab work. However, instructions and analytical processes that are fed into the computer are under the control of the human mind. Thus, it is error-prone, as in the case of sequencing data from high-throughput analysis, which often contains errors. If the sequences are wrong or annotations are incorrect, the results from the downstream analysis can be misleading as well. In sequence alignment, which in case contains errors, the structure or phylogeny related to the sequence can be ambiguous. The extent to which the computer-generated data is correct relies on the correct input and handling of data by human operators, as there is an arrangement between accuracy and computational power.

## 1.7 Conclusions

Comparison of genes using programs and tools can be a solution to the problem of having complete and reliable annotation of database sequence information with accurate biological data, as one cannot totally rely on the intelligence of this technology. DNA and protein sequences contain errors as their sources and extraction sites are infinite. In order to understand the result of a sequence in comparison with a related sequence, one needs to have the knowledge and understanding of the biology behind the sequence. The motive is to extract predictive information so as to design experiments to quickly and reliably verify the biology behind the sequence. One of the major limitations of bioinformatics is the failure of software systems in software development; these are highly costly, and so quality assurance should be considered while developing such complex software. There is a need for effective tools to perform better annotation with high accuracy and improvement of the quality of sequenced genomes without gaps. Research should be encouraged for bioinformatics tools to be more advanced for the analysis of growing high-throughput metagenomics, proteomics, and metabolomics.

## Further readings

[1]  Buehler L.K., Rashidi H.H. (Eds.). 2005. Bioinformatics Basics: Applications in Biological Science and Medicine, 2nd Edn. CRC Press. https://doi.org/10.1201/9781482292343.
[2]  Pathak R.K., Singh D.B., Singh R. Introduction to basics of bioinformatics. *Bioinformatics Methods and Applications*, 2022, 1–15. https://doi.org/10.1016/B978-0-323-89775-4.00006-7.
[3]  Verma H.N., Ghadoliya M.K., Pawar R.G., Pandit A. Bioinformatics, Universal Training Solutions Private Limited; 2013.
[4]  Raslan M.A., Raslan S.A., Shehata E.M., Mahmoud A.S., Sabri N.A. Advances in the applications of bioinformatics and chemoinformatics. *Pharmaceuticals (Basel)*, 2023, **16**(7): 1050. Published 2023 Jul 24. doi: 10.3390/ph16071050.

[5]   Xia X. Bioinformatics and Drug Discovery. *Current Topics in Medicinal Chemistry*, 2017, **17**(15): 1709–1726. doi: 10.2174/1568026617666161116143440.

[6]   Tiwari A. Applications of bioinformatics tools to combat the antibiotic resistance. In *2015 International Conference on Soft Computing Techniques and Implementations (ICSCTI)*, 2015, 96–98.

[7]   Valdivia-Granda W.A. Bioinformatics for biodefense: Challenges and opportunities. *Biosecurity and Bioterrorism: Biodefense Strategy, Practice, and Science*, 2010 Mar 1, **8**(1): 69–77.

[8]   Zhang Y., Massel K., Gao C. Applications and potential of genome editing in crop improvement. *Genome Biology*, 2019, **20**: 13.

[9]   Levy A.T., Lee K.H., Hanson T.E. *Chlorobaculum tepidum* modulates amino acid composition in response to energy availability, as revealed by a systematic exploration of the energy landscape of phototrophic sulfur oxidation. *Applied and Environmental Microbiology*, 2016, **82**(21): 6431–6439. doi: 10.1128/AEM.02111-16.

[10]  Priyadarshi M.B. Applications of bioinformatics. *Biotech Articles*, 2017. https://www.biotecharticles.com/Bioinformatics-Article/Applications-of-Bioinformatics-3270.html.

[11]  Gupta O.P., Rani S. Bioinformatics applications and tools: An overview. *CiiT International Journal of Biometrics and Bioinformatics*, 2011, **3**(3). https://www.researchgate.net/publication/264810437_Bioinformatics_Applications_and_Tools_An_Overview.

[12]  The Universal Protein Resource (UniProt). *Nucleic Acids Research*, 2008. **36**(database issue): D190–D195. doi: 10.1093/nar/gkm895.

[13]  https://www.enago.com/academy/biological-databases-an-overview-and-future-perspectives/

[14]  https://www.roseindia.net/bioinformatics/history_of_bioinformatics.shtml

[15]  https://microbenotes.com/biological-databases-types-and-importance/

[16]  https://www.ncbi.nlm.nih.gov/pubmed/24170077

---

**Multiple choice questions**

Q1   What is the primary objective of bioinformatics?
   a.   To develop new drugs and therapies
   b.   To study the structure and function of molecules
   c.   To analyze and interpret biological data using computational tools
   d.   To design and engineer new biological systems

Q2   Which of the following is not a type of bioinformatics data?
   a.   Genomic
   b.   Proteomic
   c.   Metabolomic
   d.   Anthropometric

Q3   Which of the following is not a bioinformatics tool?
   a.   BLAST
   b.   ClustalW
   c.   PCR
   d.   Phylip

Q4   Which of the following is not a type of bioinformatics database?
   a.   GenBank
   b.   UniProt
   c.   PDB
   d.   Twitter

Q5  What is the name of the file format commonly used to store DNA sequence data?
  a.  FASTA
  b.  FASTQ
  c.  SAM
  d.  BAM

Q6  What is the name of the program commonly used to align DNA sequences?
  a.  BLAST
  b.  ClustalW
  c.  MUSCLE
  d.  Phylip

Q7  Which of the following is a type of sequence alignment?
  a.  Pairwise
  b.  Multiple
  c.  Local
  d.  All of the above

Q8  What is the name of the method used to predict the 3D structure of proteins?
  a.  X-ray crystallography
  b.  NMR spectroscopy
  c.  Homology modeling
  d.  All of the above

Q9  Which of the following is a common database of protein sequences?
  a.  GenBank
  b.  UniProt
  c.  PDB
  d.  GEO

Q10  What is the name of the method used to study the function of genes by disrupting their expression?
  a.  RNA-seq
  b.  CRISPR/Cas9
  c.  Microarray
  d.  PCR

**Answers**
Q1   c.   To analyze and interpret biological data using computational tools
Q2   d.   Anthropometric
Q3   c.   PCR
Q4   d.   Twitter
Q5   a.   FASTA
Q6   b.   ClustalW
Q7   d.   All of the above
Q8   c.   Homology modeling
Q9   b.   UniProt
Q10  b.   CRISPR/Cas9

Varruchi Sharma*, Imran sheikh, Vikas Kushwaha, Shagun Gupta,
Ankur Kaushal, Seema Ramniwas, Poonam Bansal, Anupam Sharma,
Vandana Sharma, J. K. Sharma, Anil Panwar, and Anil Kumar Sharma*

# Chapter 2
# Biological databases and bioinformatics tools

**Abstract:** Modern genomic research has generated massive volumes of raw sequence data for which new computational approaches are needed to handle this enormous amount of data. Storage and effective management of such a massive amount of information have been the biggest challenges of today. However, current advancements in the field of bioinformatics have made it possible to overcome data management through the creation and utilization of computational databases. Data that is clearly defined can be accurately stored in specialized databases that adhere to ensuring rationality and consistency. Several data retrieval systems have been developed, such as Entrez, sequence retrieval system (SRS), and DBGET/LinkDB. These alignment systems not only help in providing the best and significant matches to a query but also offer valuable information from other related database sources. The genome-sequencing projects have led to the development of high-throughput technologies that generate sequence data at an incredibly fast rate. This has also resulted in the creation of computer programs that can efficiently handle and analyze huge amounts of data, extracting valuable information from it.

**Keywords:** Databases, bioinformatics tools, data management, alignment, genome sequencing

*Corresponding author: Varruchi Sharma**, Department of Biotechnology and Bioinformatics, Sri Guru Gobind Singh College, Sector 26, Chandigarh 160019, India, e-mail: sharma.varruchi@gmail.com
*Corresponding author: Anil Kumar Sharma**, Department of Biotechnology, Amity University Punjab, Mohali 140306, Punjab, India, e-mail: anibiotech18@gmail.com
**Imran sheikh**, Department of Biotechnology, Eternal University, Baru Sahib, Sirmour, Himachal Pradesh, India
**Vikas Kushwaha**, Department of Biotechnology and Bioinformatics, Sri Guru Gobind Singh College, Sector 26, Chandigarh 160019, India
**Shagun Gupta, Ankur Kaushal, Poonam Bansal**, Department of Biosciences and Technology, MMEC, Maharishi Markandeshwar (Deemed to be University), Mullana, Ambala 133207, Haryana, India
**Seema Ramniwas**, University Centre for Research and Development, University Institute of Biotechnology, Chandigarh University, Gharuan, Mohali, India
**Anupam Sharma**, Department of Physics,Guru Kashi University,Talwandi Sabo, Bathinda, Punjab, India
**Vandana Sharma, J. K. Sharma**, Department of Physics, MMEC, Maharishi Markandeshwar (Deemed to be University), Mullana, Ambala 133207, Haryana, India
**Anil Panwar**, Department of Bioinformatics and Computational Biology, College of Biotechnology, CCS Haryana Agricultural University, Hisar, Haryana, India

https://doi.org/10.1515/9783111568584-002

## 2.1 Introduction

The creation of massive volumes of raw sequence data is one of the characteristics of current genomic research. As the amount of genomic data increases, new computational approaches are needed to handle the data flood. Therefore, storing and managing such massive amounts of information become the biggest challenge that can be overcome through the creation and utilization of computational databases.

A biological database is an extensive collection of well-arranged data that is primarily maintained by computer software, which facilitates the competent data storage, retrieval, and updating varied biological information. This includes a diverse array of data types such as DNA, RNA, protein sequences, structural information, gene expression patterns, molecular interactions, mutations, phenotypic features, metabolic pathways, and taxonomic classifications of living species. Any biological database naturally designed by individual entries is added to the database one by one, with each entry in the database acting as the basic unit. These entries often contain data on nucleotides and proteins, including 3D structures, biological annotations such as gene functions, and other sequence features. They also provide information about the contributors, references, and the species from which the data was sourced [1]. Biological databases carefully ensure uniformity and logical organization, customized to efficiently handle distinct types of data. They provide accessibility, autonomy, and the capability to update content seamlessly. In addition, preestablished techniques enable the database contents to be examined efficiently. Biological databases are specifically created platforms that hold well-defined sets of data in an orderly and coherent manner. They serve as vital tools for researchers in the field. Different biological databases are summarized in Figure 2.1.

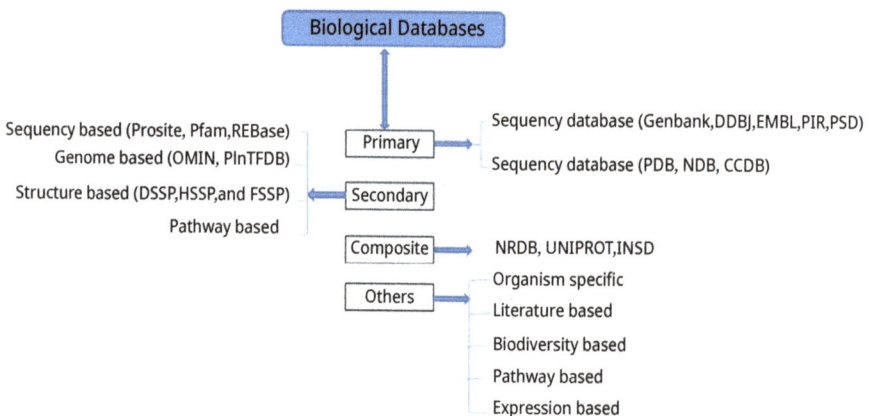

**Figure 2.1:** Classification of biological databases.

## 2.2 Classification of biological databases

### 2.2.1 Primary database

Primary databases are the most significant and major medium for research and learning purposes. The quality of the data obtained from these databases makes them special, as the data obtained is purely on a merit basis, i.e., each data is carefully curated and empirically acquired.

Each library in these databases stores natural sequencing data, as well as complete annotations, including nucleotide sequential information along with three-dimensional (3D) structural data of proteins. Every item in these databases contains essential characteristics of the corresponding sequence, offering a valuable information resource for scientists and researchers [2]. We can access these databases online via the internet from anywhere in the world, and such an accessibility makes it easy for users to explore and use important biological data. Researchers can access these databases via the extensive reach of the Internet, gaining access to a substantial amount of information and advancing their investigations. Based on its various features, the primary databases are categorized further.

#### 2.2.1.1 Sequence database

It stores DNA and protein sequence data.

##### 2.2.1.1.1 DNA/nucleotide database

DNA/nucleotide databases function as reservoirs for genetic information, containing extensive collections of DNA sequences. Every database offers distinct submission and retrieval methods specifically designed to enhance user experience. Nevertheless, a smooth and uninterrupted transfer of data takes place on a daily basis among various databases, guaranteeing consistency and conformity throughout. GenBank, EMBL, and DDBJ are prominent examples of databases that contain a vast amount of genomic data from diverse organisms. Scientists depend on these databases to obtain vital genetic data, enabling progress in disciplines such as genetics, molecular biology, and bioinformatics.

**GenBank**
The GenBank sequence database serves as a readily available repository that is carefully maintained to store nucleotide sequences and their related protein translations. Comprising a wide range of genetic information, it consists of mRNA sequences that highlight important coding regions, genomic DNA segments that include one or more genes, and clusters of ribosomal RNA genes. The GenBank database is maintained by

the National Center for Biotechnology Information (NCBI). The same database works in collaboration with the EMBL Data Library at the European Bioinformatics Institute (EBI) and the DNA Data Bank of Japan (DDBJ). This trio group (GenBank/EMBL/DDBJ) exchanges their information on a daily basis. In reference to reliable and accurate data entries, GenBank is a most trusted resource that contains huge information, such as expressed sequence tag (EST), sequence tagged site (STS), genome survey sequence (GSS), and high-throughput genome sequence data, as well as complete microbial genome sequences. In order to access information on GenBank, you can visit the server at: http://www.ncbi.nlm.nih.gov/genbank/.

### DNA Data Bank of Japan (DDBJ)

DDBJ is another major pivotal nucleotide sequence repository, which is more actively engaged in the reception of genetic data from scientific community worldwide. Once the data is submitted in the database, each dataset is accurately assigned a unique accession number that is ensuring seamless integration into the database. The database was launched in 1986 at the National Institute of Genetics, which is presently, headquartered in Mishima, Japan.

The main activities of DDBJ are as follows:

(i) In its capacity as an International Nucleotide Sequence Database (INSDC) member, DDBJ is an essential tool for researchers in their pursuit of nucleotide sequence data. Subsequently, it provides data submitters with unique accession codes and routinely shares this vital material with EMBL-Bank and GenBank.

(ii) DDBJ oversees the submission and retrieval of bioinformatics data using its tools.

(iii) Biological data analysis tools have been developed by the DDBJ.

(iv) In addition, DDBJ is also responsible for the administration of the Bioinformatics Training program that is organized in Japanese, which demonstrates researchers/students how to interpret biological data. It is also possible to obtain necessary information regarding DDBJ by utilizing the server located at http://www.ddbj.nig.ac.jp.

### European Bioinformatics Institute (EBI)

EBI is the main component of the EMBL (European Molecular Biology Laboratory). EBI was well established as EMBL-EBI and is now recognized as EMBL-Bank, which was established in 1980 in Heidelberg, Germany.

EMBL-Bank serves as a scientific hub for freely available life science data, spanning various domains of research. In the database, beyond data distribution, it incorporates fundamental research endeavors in computational biology [7]. The EMBL repository encompasses comprehensive datasets, including DNA and RNA sequences that mainly include genes, genomes, and associated variations, gene expression profiles, protein properties including sequences, families, and motifs, protein basic structural information at molecular and cellular structural levels, systems biology data constituting reactions, interactions, and data associated with pathways, chemical biology data, i.e., chemogenomics and metabolomics, and also a vast repository of scientific literature, including publications and patents. The database is accessible via the server http://www.ebi.ac.uk. EMBL-EBI stands as a cornerstone in the global scientific community [8].

### EnsEMBL

EnsEMBL is a robust software infrastructure that represents a collaborative endeavor involving the EBI, the EMBL, and the Wellcome Trust Sanger Institute. Established in 1999, it is aimed at the automatic annotation of specific eukaryotic genomes. This initiative seeks to seamlessly integrate automated annotations with existing biological data [9].

EnsEMBL's core mission encompasses automatic genome annotation, integration of this annotation with diverse biological datasets, and facilitating access to integrated information via web-based interfaces. Serving as the foundation for genome databases, EnsEMBL primarily caters to vertebrates and other eukaryotic species. Accessible over the network of computers in Asia at http://www.asia.ensembl.org, EnsEMBL serves as a pivotal resource for exploring genomic data [10].

EnsEMBL emphasizes two key domains of comparative genomics: firstly, the construction of gene trees is done, which mainly employs representative proteins from each gene within a particular species, and secondly, the alignment of DNA sequences to reveal synteny, conserved patterns, and so on [11, 12].

### 2.2.1.2 Protein database

Primary protein sequence databases (PSDs) store information on protein sequences. Here is an overview of some key PSDs.

### Protein Information Resource (PIR)

PIR (Protein Information Resource) was part of the National Biomedical Research Foundation in 1984, which had a main focus on developing tools for the identification and comprehension of protein sequence data. PIR plays a vital role in strengthening genomic and proteomic investigations and fostering scientific breakthroughs.

PIR at its core mainly comprises the PSD, housing a vast repository of over 283,000 protein sequences. For over four decades, PIR has been a stalwart provider of protein databases and analytical utilities, all freely accessible to researchers. Complementing the PSD is PIR's sophisticated bibliography system, facilitating literature searches, mapping exercises, and user submissions. Moreover, PIR extends its offerings with the Nonredundant Reference (NREF) database, which collates a wealth of protein sequence information from diverse sources, including PIR-PSD, SWISS-PROT,

TrEMBL, RefSeq, GenPept, and PDB (Protein Data Bank). This amalgamated dataset, alongside PIR's integrated database iProClass, enriches researchers with insights into protein family classification, functional attributes, and structural characteristics [12]. Currently, the PIR-NREF database encompasses more than 1,000,000 entries, providing researchers with a wealth of curated protein sequence information. The PIR, accessible at http://www.pir.georgetown.edu, serves as a gateway to these invaluable resources, empowering scientists worldwide in their quest for protein-related knowledge and discoveries.

### UniProtKB/Swiss-Prot

UniProtKB/Swiss-Prot represents the meticulously curated and peer-reviewed segment within the UniProt knowledgebase (UniProtKB), offering a nonredundant repository of annotated protein sequences. Originating in 1986, Swiss-Prot is a collaborative effort between the EMBL Outstation (EBI) and the Swiss Institute of Bioinformatics (SIB), serving as a gold standard in PSDs. Distinguished by its manual annotation process, Swiss-Prot provides comprehensive information on protein attributes, including domain structure, function, posttranslational modifications, and variants. It stands apart from other PSDs by virtue of three key criteria: meticulous annotation, minimal redundancy, and seamless integration with complementary databases.

A translation of the EMBL nucleotide sequence database, known as TrEMBL, was added to Swiss-Prot in 1996. The computer-annotated supplement TrEMBL was added to the EMBL database, which includes automatically annotated entries that are obtained from the translation of all coding sequences (CDSs) inside the database, with the exception of those that are already annotated in Swiss-Prot [10, 14, 58]. The main servers hosting SwissProt and TrEMBL are accessible at www.ebi.ac.uk/uniprot and www.ebi.ac.uk.

### Protein sequence database (PSD)

PIR-International PSD (PIR-PSD) stands as a pioneering achievement in the realm of PSDs, tracing its origins to the landmark Atlas of Protein Sequence and Structure (1965–1978), methodically curated under the guidance of Margaret Dayhoff. This groundbreaking endeavor laid the foundation for PIR-PSD, the world's premier repository of classified and functionally annotated protein sequences [31].

The database is accessible at https://proteininformationresource.org/.

**Structure database:** It stores the 3D shapes of biological macromolecules like proteins and nucleic acids. Some important structure databases are given below:

### Protein Data Bank (PDB)

In 1971, the PDB was founded at Brookhaven National Laboratory. It serves as a repository for 3D structures of proteins, elucidated through X-ray crystallographic and nuclear magnetic resonance studies. It is managed by the Research Collaboratory for Structural Bioinformatics at Rutgers University.

The database holds a number of protein structures. These structures provide detailed insights into the atomic coordinates of amino acids within proteins, including protein fragments and proteins bound to substrates or inhibitors. Depositing protein structure data into PDB is facilitated by the web-based AutoDep Input Tool, streamlining the submission process for researchers. Once deposited, the molecular structures within PDB can be visualized using molecular graphics programs such as ResNet and Chine CnED, enabling researchers to explore and analyze protein structures with precision. The database can be accessed through http://www.rcsb.org/pdb/ [27].

### Nucleic Acid Database (NDB)

The 3D structures of nucleic acids can be obtained from the Nucleic Acid Database (NDB). In the year 1992, the database was established. The main aim of the NDB project was to collect, organize, and share information about nucleic acid structures. Wilma K. Olson and Helen M. Berman of Rutgers University and David Beveridge of Wesleyan University were the founders of the project. The database can be accessed at http://ndbserver.rutgers.edu and http://www.ebi.ac.uk/NDB/.

The Cambridge Structural Database (CSD), also known as the Cambridge Crystallographic Data Center (CCDC), is a valuable resource in the field of biotechnology. The CCDC is an esteemed educational institution specializing in the fascinating field of crystallography. It is situated in the historic city of Cambridge in England. This process generates and distributes the CSD. CSD maintains a database of small molecules. Experimental data on organic and metal-organic crystal structures are stored in that database. The website for CSD is www.ccdc.cam.ac.uk.

## 2.2.2 Secondary database

The secondary database comprehends data analyzed from primary sequence data. The data derived is presented in various formats such as regular patterns, fingerprints, blocks, profiles, or hidden Markov models (HMMs). Some of the major secondary databases include ProSite, Profiles, Prints, Pfam, and REBase, which are categorized into four main groups: sequence-related data, genome-related data, structure-related data, and pathway databases. Each of these categories provides a unique perspective on the analysis and interpretation of biological data, contributing to a comprehensive understanding of sequence information and its functional implications.

### 2.2.2.1 Sequence-related database

**ProSite**

ProSite is a prominent secondary protein database, which provides detailed annotations pertaining to protein families, domains, functional sites, as well as amino acid patterns and profiles. This curated information in the database is accurately assembled by the SIB, ensuring high-quality data integrity. ProRule is the significant complement to the ProSite database, which extends upon the domain descriptions provided by ProSite. ProRule enriches the dataset by furnishing critical insights into structurally and functionally significant amino acids. The database is accessible via http://www.prosite.expasy.org.

**Pfam**

Pfam database consists of an extensive collection of protein families, which offers comprehensive resources such as multiple sequence alignments (MSAs) and HMMs for protein domains across diverse organisms. Pfam systematically generates information on MSAs, enabling users to explore protein domain structures, analyze species distribution patterns, cross-reference with other databases, and visualize known protein structures [5].

Pfam mainly comprises two primary components: Pfam-A and Pfam-B. Pfam-A consists of manually curated entries of the highest quality; each entry is an accurately curated protein sequence alignment and HMM. On the other hand, Pfam-B holds automatically generated entries of lower quality. While Pfam-A entries represent a subset of known proteins, Pfam-B families offer a broader scope, catering to instances where comprehensive coverage is required [28].

**ExplorEnz**

ExplorEnz is an open-access database created and maintained by Andrew McDonald. It provides access to a vast amount of data related to the enzyme nomenclature list of the International Union of Biochemistry and Molecular Biology. Explore a vast collection of rigorously scrutinized and verified information about enzymes, which serves as a dependable resource for researchers and enthusiasts to delve into the captivating field of enzyme biochemistry.

**REBase**

REBase is a comprehensive database that is used to collect and rigorously organize information on restriction enzymes and their associated proteins. REBase has a vast amount of information, such as recognition and cleavage sites, isoschizomers, availability for purchase, sensitivity to methylation, crystallographic data, genomic information, and sequence data. Explore the complex realm of DNA methyltransferases, homing endonucleases, nicking enzymes, specificity subunits, and regulatory proteins, all within the extensive boundaries of REBase, which is accessed at http://www.rebase.neb.com [36].

## 2.2.2.2 Genome-related database

### Online Mendelian inheritance in man (OMIM)
OMIM (Online Mendelian Inheritance in Man) serves as a comprehensive repository, highlighting major information on human genes, genetic disorders, and genetic variations. OMIM offers references for further research and provides tools for genomic analysis of the genes cataloged within its database. The database was started in the early 1960s in the form of a book. Its transition to an online platform occurred in 1987 with the inception of OMIM. Data collection and processing for OMIM are conducted at Johns Hopkins University [26].

A six-digit number is used to represent data in a database, with the first digit used to classify the inheritance mechanism. For example, a characteristic that is "1" indicates an X-linked phenotype, "2" an autosomal recessive trait, and "3" an autosomal dominant trait.

Various symbols precede the entry numbers to convey specific information: a number symbol (#) indicates a descriptive entry, a plus sign (+) indicates an entry containing the description of a gene with a known sequence and phenotype, a percent sign (%) indicates an entry describing a confirmed Mendelian phenotype or phenotypic locus with an unknown molecular basis, and no symbol before an entry number indicates a description of a phenotype without a clearly established Mendelian basis. Furthermore, a caret (^) before an entry number denotes that the entry has been removed from the database or relocated to another entry.

The database can be accessed at http://www.omim.org.

### Plant Transcription Factor Database (PlnTFDB)
Plant Transcription Factor Database (PlnTFDB) serves as a publicly accessible database dedicated to cataloging plant genes involved in transcriptional control. Operating via its web portal (http://plntfdb.bio.uni-potsdam.de/v3.0/), PlnTFDB offers an integrated repository housing comprehensive collections of transcription factors (TFs) and other transcriptional regulators (TRs). These datasets are derived from plant species with fully sequenced and annotated genomes. Spanning across 19 species, ranging from unicellular red and green algae to angiosperms, PlnTFDB encompasses complete sets of 84 TF and TR families, representing over 1.6 billion years of evolutionary history in gene regulatory networks. Each entry for TF or TR genes includes detailed information such as ESTs, 3D protein structures of homologous proteins, domain architecture, and cross-references to other computational resources available online. Additionally, PlnTFDB

facilitates cross-species comparisons by linking different species through orthologous genes, fostering enhanced understanding and exploration of plant TR mechanisms.

### 2.2.2.3 Databases pertaining to structure

The DSSP database is a valuable resource that provides assignments for secondary structure.

Standardization of secondary structures is performed using the DSSP tool, which also stores the secondary structure assignments for all proteins included in the PDB [27]. However, the same tool does not have the capability to predict secondary structures of proteins.

Database of homology-derived secondary structures of proteins is explored. It contains secondary structure, sequence variability, aligned sequence, and sequence profile for all known protein structures in the database. This also provides insight into a protein's tertiary structure.

Protein families with shared structures (FSSP – families of structurally similar proteins) are explored. The structurally similar proteins are stored in the "distance matrix alignment" (DALI) database, also known as FSSP. The database is much more significant for protein structural comparisons. DALI archives according to the comparisons of 3D protein structures in PDB. The DALI database serves as the secondary structure database.

### 2.2.2.4 Database of information on pathways

Pathway databases contain valuable information on genes, proteins, chemical substances, and their reactions, along with molecular interactions in biological processes.

## 2.2.3 Composite database

The composite databases gather data from different primary databases. This is demonstrated by the following examples of composite databases:

GenBank (CDS translations), PDB, Swiss-Prot, PIR, and PRF are all sources of sequences that are merged and stored in nonredundant database. The nucleotide sequences from EMBL, GenBank, and DDBJ are all stored in INSD. UniProt, consisting of the Universal PSD, Swiss-Prot, and PIR-PSD, gathers protein sequences from various locations across the globe.

## 2.2.4 Additional databases

Aside from databases that rely on literature, there are also databases dedicated to specific organisms and biodiversity informatics. Organism-specific databases contain a diverse range of nonhuman creatures, such as viruses, bacteria, fungi, invertebrates, beetles, silkworms, and *Drosophila*. Organism-specific databases store a wide range of information, including genome maps, gene expression data, genetic mutations in *Drosophila*, morphological traits of yeast, and growth needs of bacterial pathogens [43–45]. Scientific articles published by researchers in various journals can be accessed through literature-based databases, either as abstracts or full papers. Among various databases that compile scholarly articles, PubMed is the most popular and widely used. There is a vast amount of data stored in databases that are specifically designed for biodiversity informatics, which focuses on the wide range of life forms found on our planet.

A list of different biological databases, along with their sources, is summarized in Table 2.1.

**Table 2.1:** Frequently used essential biological databases.

| S. no. | Database | Source/objective | URL | References |
|--------|----------|------------------|-----|------------|
| 1 | CATH | Delivers data on the evolutionary relationships of protein domains | https://www.cathdb.info/ | [5] |
| 2 | ChIPBase | Long noncoding RNA and microRNA gene regulatory database | http://deepbase.sysu.edu.cn/chipbase | [6] |
| 3 | dbSNP | Database of single nucleotide polymorphisms | http://www.ncbi.nlm.nih.gov/snp | [7] |
| 4 | DIP | Database of Interacting Proteins | http://dip.doe-mbi.ucla.edu | |
| 5 | DDBJ | National Institute of Genetics/DNA sequences | | [9] |
| 6 | DisProt | Database of fundamentally disordered proteins | https://www.disprot.org/ | [9] |
| 7 | EcoCyc | *E. coli* database | https://ecocyc.org/ | [10] |
| 8 | EMBL | European Bioinformatics Institute | https://www.embl.org/ | [11] |
| 9 | EnsEMBL | Genome annotation databases are automatically updated for mammals, birds, and other eukaryotic organisms | https://asia.ensembl.org/ | [12] |
| 10 | FlyBase | Drosophila | https://flybase.org/ | [13] |

**Table 2.1** (continued)

| S. no. | Database | Source/objective | URL | References |
|---|---|---|---|---|
| 11 | GenBank | National Center for Biotechnology Information/Nucleotide Sequence Database | https://www.ncbi.nlm.nih.gov/genbank/ | [14] |
| 12 | GENCODE | A gene and its variants: encyclopedia | http://www.gencodegenes.org | [15] |
| 13 | GeneCards | An integrated database of human genes | http://www.genecards.org | [16] |
| 14 | GO | Gene ontology | http://geneontology.org | [17] |
| 15 | InterPro | Provides the functional exploration of proteins | https://www.ebi.ac.uk/interpro/ | [18] |
| 16 | KEGG | Kyoto Encyclopedia of Genes and Genomes | http://www.genome.jp/kegg | [19] |
| 17 | KEGG Pathway | KEGG pathway maps | http://www.genome.jp/kegg/pathway.html | [20, 21] |
| 18 | Legume Information System (LIS) | Genomic database for the legume family | https://legumeinfo.org/ | [22] |
| 19 | MetaCyc | Metabolic pathway database | http://metacyc.org | [23] |
| 20 | NCBI RefSeq | NCBI Reference Sequence Database | http://www.ncbi.nlm.nih.gov/refseq | [24] |
| 21 | NONCODE | Database of ncRNA genes | http://www.noncode.org | [25] |
| 22 | OMIM | Human genes and genetic phenotypes database | https://www.ncbi.nlm.nih.gov/omim/ | [26] |
| 23 | PDB | 3D structure of proteins and nucleic acids | https://www.rcsb.org/ | [27] |
| 24 | Pfam | Protein families database of alignments and HMMs | https://pfam.xfam.org/ | [28] |
| 25 | PHI-base | Molecular and biological data on genes influence the results of interactions between pathogens and hosts | https://bio.tools/phi-base | [29] |
| 26 | PID | Pathway Interaction Database | http://pid.nci.nih.gov | [30] |
| 27 | PIR | Protein Information Resource | http://pir.georgetown.edu | [31] |

**Table 2.1** (continued)

| S. no. | Database | Source/objective | URL | References |
|---|---|---|---|---|
| 28 | PomBase | Yeast *Schizosaccharomyces pombe* | https://www.pombase. org/ | [32] |
| 29 | PRINTS | A collection of protein fingerprints | http:// 130.88.97.239/PRINTS/ index.php | [33] |
| 30 | PROSITE | Database of protein families and domains | http://www.expasy. org/prosite | [34] |
| 31 | PUBMED | Database of literature | https://www.ncbi.nlm. nih.gov/pubmed/ | [35] |
| 32 | PubMed Central | Literature archive | http://www.ncbi.nlm. nih.gov/pmc | [36] |
| 33 | Rfam | Database of ncRNA families | http://rfam.xfam.org | [37] |
| 34 | Rat Genome Database | Genomic and phenotype data for *Rattus norvegicus* | https://rgd.mcw.edu/ | [38] |
| 35 | RefSeq | A repository of selected and annotated sets of nucleotide sequences and protein products | https://www.ncbi.nlm. nih.gov/refseq/ | [39] |
| 36 | Saccharomyces Genome Database | Genome of the yeast model organism | https://www.yeastge nome.org/ | [40] |
| 37 | SCOP | Structural classification of proteins | http://scop.mrc-lmb. cam.ac.uk/ | [41] |
| 38 | UCSC Genome Browser | UCSC Genome Browser database | http://genome.ucsc. edu | |
| 39 | UniProt | A collection of protein sequence databases that includes Protein Information Resource, Swiss-Prot, and TrEMBL | https://www.uniprot. org/ | [42] |
| 40 | VectorBase | Provides genetic, phenotypic, and population-centric data for invertebrate vectors of human pathogens | https://www.vector base.org/ | [43] |
| 41 | WormBase | For the nematodes *Caenorhabditis elegans* | https://wormbase. org//#012-34-5 | [44] |
| 42 | Xenbase | Genome of *Xenopus tropicalis* and *Xenopus laevis* | http://www.xenbase. org/entry/ | [45] |

## 2.3 Biological database retrieval system

With Entrez at NCBI [46], the SRS at EBI [47], and DBGET/LinkDB [48] in Japan, there are three major data retrieval systems. These retrieval systems not only deliver matches to a query but also offer valuable supplementary information in related databases.

### 2.3.1 Entrez

Introducing Entrez, a sophisticated database and retrieval system developed by the prestigious National Center for Biotechnology Information (NCBI). Accessible via the website http://www.ncbi.nlm.nih.gov/Entrez/, Entrez provides access to a wide range of highly valuable resources. Explore the extensive nucleotide sequence databases, such as GenBank/DDBJ/EBI, as well as PSDs such as Swiss-Prot, PIR, PRF, PDB, and translated protein sequences. Access a comprehensive literature collection, including PubMed, which provides access to MEDLINE and pre-MEDLINE papers. Additionally, explore genome and chromosome mapping data as well as molecular modeling of 3D structures.

Explore the complex realm of taxonomy by gaining access to DNA and protein sequences for diverse taxonomic categories, and explore specialized databases such as OMIM, dbSNP, and UniSTS. Entrez utilizes the notion of "neighboring" to provide effortless access to similar articles in linked databases, hence improving the study discovery. Utilize the Entrez system to efficiently extract extensive datasets that meet particular criteria and effortlessly transfer them to your local PC. Entrez is an essential tool for molecular biologists and researchers because of its easy-to-use interface and extensive range of capabilities. It facilitates revolutionary discoveries in the field of biology (Table 2.2) [49].

**Table 2.2:** Salient features of the Entrez database.

| S. no. | Name of database | Objective of database | URL |
|---|---|---|---|
| 1 | Assembly | Using the assembly database, we can approach the genomic assemblies for either the acquired data or NCBI RefSeq assemblies | https://www.ncbi.nlm.nih.gov/assembly |
| 2 | BioProject | It is an assemblage of complete or in-progress large-scale molecular projects | https://www.ncbi.nlm.nih.gov/bioproject/ |
| 3 | BioSample | Provides biological source data used in studies | https://www.ncbi.nlm.nih.gov/biosample |
| 4 | BioSystems | Information on interrelating sets of biomolecules | https://www.ncbi.nlm.nih.gov/biosystems/ |

**Table 2.2** (continued)

| S. no. | Name of database | Objective of database | URL |
|---|---|---|---|
| 5 | Bookshelf | Full-text books accessible online | https://www.ncbi.nlm.nih.gov/books/ |
| 6 | ClinVar | An archive of reports of clinically relevant human genetic variations | https://www.ncbi.nlm.nih.gov/clinvar/ |
| 7 | Conserved Domains | Provides information on domains that are conserved in molecular evolution | https://www.ncbi.nlm.nih.gov/cdd |
| 8 | dbGaP | Database of Genotypes and Phenotypes | https://www.ncbi.nlm.nih.gov/gap/ |
| 9 | dbVAR | Database of Genomic Structural Variation | https://www.ncbi.nlm.nih.gov/dbvar/ |
| 10 | EST | Expressed sequence tag | https://www.ncbi.nlm.nih.gov/nuccore/ |
| 11 | Gene | Exploratory database of genes | https://www.ncbi.nlm.nih.gov/gene |
| 12 | Genome | Encompasses sequence and map data from whole genomes | https://www.ncbi.nlm.nih.gov/genome/ |
| 13 | GEO Datasets | Depository of microarray data | https://www.ncbi.nlm.nih.gov/gds |
| 14 | GEO Profiles | Storehouse of discrete gene expression and molecular profusion profiles | https://www.ncbi.nlm.nih.gov/geoprofiles?db=geo |
| 15 | GSS | Comprehends sequence records from the bulk GSS | https://www.ncbi.nlm.nih.gov/nuccore/ |
| 16 | GTR | Genetic Testing Registry | https://www.ncbi.nlm.nih.gov/gtr/ |
| 17 | HomoloGene | Comprises homologous genes and corresponding mRNA | https://www.ncbi.nlm.nih.gov/homologene |
| 18 | MedGen | Portal of human disorders | https://www.ncbi.nlm.nih.gov/medgen |
| 19 | MeSH | Medical Subject Headings are used for indexing articles in PubMed | https://www.ncbi.nlm.nih.gov/mesh |
| 20 | NCBI Web Site Search | NCBI web pages, certification, and available online tools | https://www.ncbi.nlm.nih.gov/ncbisearch/ |
| 21 | NLM Catalog | Comprehends books, journals, audiovisuals, software, electronic means of information, etc., in the assemblages of NLM | https://www.ncbi.nlm.nih.gov/nlmcatalog/ |

**Table 2.2** (continued)

| S. no. | Name of database | Objective of database | URL |
|---|---|---|---|
| 22 | Nucleotide | All sequence data as of GenBank, EMBL, and DDBJ | https://www.ncbi.nlm.nih.gov/nuccore |
| 23 | OMIM | Online Mendelian Inheritance in Man | https://www.ncbi.nlm.nih.gov/omim |
| 24 | PopSet | The same covers interrelated nucleotide sequences | https://www.ncbi.nlm.nih.gov/popset/ |
| 25 | Probe | Database of nucleic acid reagents | https://www.ncbi.nlm.nih.gov/probe/ |
| 26 | Protein | Contains amino acid sequences | https://www.ncbi.nlm.nih.gov/protein/ |
| 27 | Protein Clusters | Admittance to annotation information, books, journals, domains, assemblies, and analysis tools, etc. | https://www.ncbi.nlm.nih.gov/proteinclusters/ |
| 28 | PubChem BioAssay | PubChem BioAssay is a comprehensive database that offers bioactivity screens of chemical substances found in PubChem Substance | https://www.ncbi.nlm.nih.gov/pcassay/ |
| 29 | PubChem Compound | The PubChem Compound database provides distinctive, authenticated chemical structure details | https://www.ncbi.nlm.nih.gov/pccompound/ |
| 30 | PubChem Substance | Provides statistical data on chemical substances submitted to PubChem by researchers | https://www.ncbi.nlm.nih.gov/pcsubstance/ |
| 31 | PubMed | Database of biomedical literature | https://pubmed.ncbi.nlm.nih.gov/ |
| 32 | PubMed Central | A library of life sciences journal literature | https://www.ncbi.nlm.nih.gov/pmc/ |
| 33 | SNP | Single nucleotide polymorphism | https://www.ncbi.nlm.nih.gov/snp/ |
| 34 | SRA | Sequence Read Archive | https://www.ncbi.nlm.nih.gov/sra |
| 35 | Structure | Molecular Modeling Database (MMDB) comprehends investigational data from X-ray crystallographic and nuclear magnetic resonance grids | https://www.ncbi.nlm.nih.gov/Structure/index.shtml |
| 36 | Taxonomy | Contains phylogenetic lineages of different species | https://www.ncbi.nlm.nih.gov/taxonomy/ |
| 37 | UniGene | It provides information on protein relationships, gene expression, cDNA clone reagents, and genomic locations, etc. | https://www.ncbi.nlm.nih.gov/unigene |

## 2.3.2 Sequence retrieval system (SRS)

SRS functions as a network browser for databases [47] (http://srs.ebi.ac.uk/). The EBI in Hinxton, United Kingdom, is responsible for developing 80 biological databases, which are searched by SRS. Databases pertaining to sequences and sequence-related information, metabolic pathways, TFs, application results (such as BLAST), protein 3D structures, genomes, mapping, mutations, and locus-specific mutations are all accessible through this interface. One of the benefits of SRS is that it searches queries in a very short amount of time, which enables users to obtain, link, and access information from different resources that are related to one another.

## 2.3.3 DBGET/LinkDB

This is GenomeNet's bioinformatics database retrieval system, created by Kyoto University's Institute for Chemical Research and the University of Tokyo's Human Genome Center [48]. Accessed through http://www.genome.ad.jp/dbget/, DBGET provides access to around 20 molecular biology databases that can be searched individually. After making a query in one of the databases, DBGET conveniently offers links to related information alongside the list of results. DBGET has a unique partnership with the Kyoto Encyclopedia of Genes and Genomes (KEGG) database, which houses valuable data on metabolic and regulatory pathways. The data provided includes information on protein relationships, gene expression, cDNA clone reagents, and genomic location, among other things.

# 2.4 Bioinformatics tools

The genome-sequencing projects led to the development of high-throughput technologies, resulting in the rapid production of sequence data. The task involved developing computer programs that could efficiently handle and organize vast amounts of data, extracting the required information from the stored data [50–60]. Due to these circumstances, significant enhancements were made to computer hardware capabilities, novel statistical techniques were devised, meticulous planning was done for computer programs, and suitable data storage and management systems were implemented. The different computer programs utilized for data acquisition, analysis, correlation identification, pattern recognition, and achieving specific goals are commonly known as bioinformatics tools.

A large range of different resources are available for gene prediction, interaction analyses, data storage, and management, which are briefly described below.

**AutoSNP:** Nucleotide fragments with expressed gene tags can be studied to identify single nucleotide polymorphisms (SNPs). These SNPs have great biological significance as they are based on expressed gene exons. AutoSNP computer software conducts automatic analysis of EST sequence data and detects SNPs as well as combinations of insertion/deletion (InDel) in them. This aligns the EST sequences and distinguishes between predicted SNPs and redundancy criterion sequencing errors. A putative SNP appears in several reads, while a sequencing error occurs in one or two reads. Redundancy and co-segregation scores are calculated for each SNP nominee. A projected SNP locus redundancy score is the polymorphism frequency at this locus. The co-segregation value is the probability that the expected SNP will be distributed in the EST series with the other SNPs in its vicinity. The AutoSNP performance includes the SNPs and InDels with their consistency and co-segregation ratings. Some SNPs and InDels predicted using the AutoSNP method in maize were validated as true SNPs and InDels. The AutoSNP software is open to research staff free of charge on request [61]. The SNPServer (http://hornbill.cspp.latrobe.edu.au/snpdiscovery.html) is a web interface for real-time AutoSNP, BLAST, and CAP3 discovery programs [62]. BLAST recognizes similar EST sequences, CAP3 aligns and clusters these sequences, while AutoSNP analyzes SNP and InDel alignments. The results of this SNP discovery process, the EST data source, and their annotation are stored in autoSNPdb. This database can be accessed at http://autosnpdb.qfab.org.au/. The database has rice, barley, and *Brassica* species SNP data. AutoSNPdb helps to classify SNPs and InDels in specified genes or genes linked to different traits and between genes of specified pairs/plant varieties. A user-friendly GUI (graphical user interface) makes a simple visualization of the database SNPs [63]. One tool, Quality SNPng, uses a haplotype-based strategy to allow NGS data to visualize and detect SNPs and does not require a completely sequenced reference genome (http://www.bioinformatics.nl/QualitySNPng/).

**SNP2CAPS:** CAPS markers are useful, cost-effective methods for studying polymorphisms in laboratories that are not highly equipped. This is particularly valid when using common restriction enzymes to examine CAPS markers. Converting SNP markers into CAPS markers is very difficult. The dCAPS Finder 2.0 computer program can be used to design PCR primers with these mismatches that either create a restriction site at the selected SNP locus or delete a locus site. This facilitates converting SNPs to CAPS markers, but successfully developing these primers is not a simple task. The SNP2CAPS software screens multiple aligned sequences for polymorphic restriction sites, analyzes these sites, and identifies those sites that are most likely candidates for CAPS marker development. This standardized software also tests the restriction enzymes in the submitted sequences for their suitability for CAPS analysis and selects enzymes showing at least one restriction site polymorphism in each associated sequence [64]. If polymorphism at a restriction site occurs from an appropriate SNP, the restriction site should have an unambiguous sequence, i.e., consisting only of A, T, C, and G. However, in some cases, uncertainty (symbol N) may occur in one or two positions in addition to SNP polymorphism, but this is unlikely to affect CAPS development. The above cases would be good candidates for CAPS development. Yet restriction site polymorphisms can arise merely due to an unclear sequence, in which case N may occur in any of the associated sequences within the restriction site series. These site-restricted polymorphisms are not appropriate for CAPS. Alternatively, inserting (or deleting) one or more nucleotides into (or from) the restriction site will also produce restriction site polymorphism, which would be useful for CAPS development. Therefore, the SNP2CAPS software analyzes the submitted sequences and identifies polymorphisms suitable for CAPS marker development. SNP2CAPS system input is an MSA of target sequences from various accessions. This can be in modified FASTA, ClustalW, MSF, MEME, and many other formats. It also requires a restriction enzyme data input that can be downloaded from REBASE (restriction enzyme database; http://rebase.neb.com/). A high proportion of ~90% of multiple aligned barley EST sequences contained SNPs and InDels, and one or more polymorphic restriction sites. However, over 30% of these polymorphic restriction sites were for 10 common enzymes. SNP2CAPS provides a command line and GUI. The SNP2CAPS is downloadable from http://pgrc.ipk-gatersleben.de/snp2caps/.

**TASSEL (Trait Analysis by aSSociation, Evolution, and Linkage):** Factors such as selection, population composition, and family relationships sometimes complicate the findings of association analyses, leading to incorrect marker–trait associations. GLM (general linear model) and MLM (mixed linear model) methods were developed to mitigate the impact of population structure and/or family relationships in association with study findings. GLM and MLM approaches were introduced by aSSociation, Evolution, and Linkage in the TASSEL program. The GLM approach uses a hierarchical $Q$-matrix-based association analysis to minimize false association likelihood. The $Q$ matrix represents population structure and is determined using either the STRUCTURE system or the PCA (principal component analysis) process. In its model, the MLM approach uses the kinship ($K$) matrix and the $Q$ matrix to further the chance of finding false-positive correlations. $K$ matrix estimates describing the average relationship between pairs of individuals/lines can be derived from pedigree or genotype data for a large number of unlinked markers covering the organism's entire genome. TASSEL performs F-checks, permutation checks, and model impact estimates. If the trait concerned does not usually have a distributed residual error, any transformation function may be used to produce approximately normal error terms, or a permutation method may be used to produce distribution-independent $P$ values. The TASSEL software can manage plant, animal, and human datasets. This allows the calculation of linkage imbalance (LD) as $D'$ and $r^2$ and allows the graphical representation of these figures. Other features of this software include analyzing InDels, analyzing diversity, executing PCA, and imputing missing data. This package includes many data extraction and visualization tools, including sequence alignment viewer, neighbor-joining cladogram creation, and many data graphing functions. This has several data management features and a web explorer that offers an interface to relational databases. Java software is compatible with Windows, Mac, and

Linux operating systems. TASSEL executables, user manuals, and so on are freely accessible from http://www.maizegenetics.net/tassel [65].

**Structure:** Using STRUCTURE software, the analysis of two or more similar subsets within a larger population can be easily identified [66]. A homogeneous group is a group of individuals at Hardy–Weinberg equilibrium for all the random markers. This program uses a Bayesian (Markov chain Monte Carlo) approach to group people based on models for different, unrelated indicators. Information from SSRs, SNPs, and AFLPs, among other genetic markers, may be used. The main goal is to determine how many demographic subsets are most likely to exist in the population. Also, based on the population marker data, the investigator will attempt to determine the minimum number of groups that adequately depict the basic structure. Additionally, it generates $Q$ estimates, which display the likelihood that a person is a member of a certain cluster. It is possible to classify an individual into more than one category if their $Q$ scores indicate that they share certain hereditary traits. The precision of these assignments relies on various factors, such as the number of individuals genotyped in the sample, the groups represented in the sample, the marker loci analyzed, the level of admixture in the population, and the extent of variations in the sample allele frequency. This software was utilized to analyze genetic structures in sampled populations, categorize individuals into specific sample groups, conduct population admixture analysis, and examine hybridization patterns, among other applications. Research indicates that STRUCTURE has proven to be highly effective in accurately assigning individuals to their respective origin populations, especially in cases where the population consists of two to four distinct and homogeneous groups. Once the STRUCTURE software is initiated with a random configuration, it undergoes a series of simulation steps, typically ranging from 10,000 to 100,000. Additional steps, often exceeding 10,000–100,000, are then taken to obtain precise $Q$ estimates. The program runs multiple iterations on the dataset, testing different numbers of classes, ranging from 1 to 10. The program executables are compatible with Mac, Windows, Linux, or Sun operating systems. The program's computational component is written in C and has a Java front end, which offers a wide range of useful features. The data file should be a text file, and a number, typically 9, will indicate the missing data which is not used elsewhere in the data file. The STRUCTURE software is accessible at http://pritch.bsd.uchicago.edu/structure21.html.

**Microarray software:** An advanced experimental approach for determining patterns and rates of gene expression throughout the whole genome is known as microarray technology. It generates massive amounts of data that require intuitive software for gathering, analyzing, storing, and managing.

The TM4 program is a suite of four tools: (i) A MicroArray DAta Manager (MADAM) tool helps in RNA isolation and data processing using this technique. It provides a platform for the deployment of additional data analysis and management capabilities and promotes database entry. (ii) The TIGR Spotfinder method analyzes microarray images quickly and reproducibly and quantifies gene expression rates. (iii) The MIcroarray

Data Analysis System (MIDAS) normalizes and filters Spotfinder-generated data. Finally, (iv) the MultiExperiment Viewer (MeV) tool analyzes data files on gene expression and shows gene expression and annotation information from microarray experiments. PCA, hierarchical and k-means clustering, self-organizing maps and trees, and so on are all part of the MeV study modules. Consensus clusters are established via bootstrapping and jackknifing. In addition, TM4 stores its data in a MySQL database. The MIAME standards for limited details on a microarray experiment are met by this database. Although TM4 was first developed for spotted two-color microarrays, it is readily adaptable to single-color microarray formats. All sorts of living things, from plants to animals to microbes, may benefit from TM4. We can check out this open-source software package (http://www.tigr.org/software). Also, various operating systems are compatible with MADAM, MIDAS, and MeV, but TIGR Spotfinder is Windows only [67].

**MAPMAN:** It generates massive amounts of data spanning a wide range of factors via analysis utilizing microarrays and mass spectrometry-based metabolite profiling. Meticulous examination and interpretation are required to meet the growing challenge. The MAPMAN device allows for the examination of large datasets by displaying them as diagrams depicting metabolic pathways or other relevant biological activities and processes. The SCAVENGER and IMAGEANNOTATOR are the modules that make up this approach [68].

The SCAVENGER module sorts metabolite and gene expression data into "Bins" and "subBins," which are the two levels of a hierarchical classification system. One area of metabolism, like photosynthesis, is associated with a Bin. Examples of subBins inside a Bin are "light reactions," "photorespiration," and the "Calvin cycle" for "photosynthesis." Bins and subBins are assigned similar integer codes to illustrate their hierarchical links. Gene expression data and metabolite data are handled by separate SCAVENGER modules. Annotated gene roles are used to classify genes into Bins and subBins by the TRANSCRIPTSCAVENGER module, which processes data from gene expression arrays. Bin and subBin data assignment involves both automated recruiting and human adjustment.

The assignment guidelines are as follows: (1) All genes, including those with "supposed" annotation, should ideally be placed in separate bins, (2) bin structure should be changed to fit relevant data, and (3) as far as possible, each gene should be put in a single bin and subBin.

The structures and routes of the metabolites are the basis for the METABOLITESCAVENGER's classification system. The data groups generated by SCAVENGER are organized and shown as diagrams by IMAGEANNOTATOR. The modular design of MAPMAN makes it easy to add novel information categories, change current ones, and even build SCAVENGER modules for different types of data. MAPMAN needs to be enhanced in order to fix its limitations and include more applications.

In order to automate the process of updating annotation and terminology via error-free acquisition of the Gene Ontology Consortium (GOC) and associated releases,

SCAVENGER modules must be developed. Additionally, modules are needed to elimi-
nate redundant data and provide absolute quantities of gene expression or metabolite
accumulation. Modules that can analyze statistical data also need to be developed.

The IMAGEANNOTATOR module is available through http://gabi.rzpd.de/projects/
MapMan/.

**GenScan:** The prediction of introns/exons/complete gene structures, exon–intron borders,
promoter sites, and poly-A signals in genome sequences of various organisms is done
through GenScan for several different organism types [69]. In this program, the gene
structure model is a "probabilistic approach" mainly for human genes, which includes a
detailed explanation of transcription, translation, and splicing signals. This may also include features related to exons/introns, intergenic region lengths, and base compositions.
It scans the query sequence for this model's features, identifies the stretches of the sequence matching exons, promoters, etc., descriptions, and assigns a probability to each
defined stretch. The defined "optimal exons" suit the highest probability model ($P > 0.99$)
and are assumed to represent actual exons. Furthermore, GenScan predicts "suboptimal
exons" with appropriate likelihood rates representing a true exon. Exons with <0.50 likelihood are discarded as unreliable. This software can predict multiple genes and partial
genes in a given nucleotide series. Users will analyze the "optimal" and "suboptimal" prediction sets to distinguish nonstandard gene structures including spliced genes. GenScan
can recognize and evaluate nucleotide sequences of up to 1 million base pairs. We can
study the sequence of one or both DNA duplex strands and reliably predict gene classes.
GenScan is highly accurate but sensitive to exon length. GenScan is the most detailed and
advanced tool available for free. You can access the GenScan server at http://genes.mit.
edu/GENSCAN.html. Other methods are FGENESH/FGENES, HMM Gene, GENE ID, GENE
PARSER, and so on.

**ClustalW:** MSA is possibly the most effective bioinformatics investigation procedure. It
has proven itself as a most beneficial program in almost every case. It helps in predicting protein structure/function, which is the basis of phylogenetic analysis. The Clustal
system family is the most commonly used to coordinate multiple sequences at a time.
There are two types of Clustal programs (ver. 2): (i) ClustalW (which is a command line
user interface) and (ii) ClustalX (a GUI) [70–72]. ClustalW is the most frequently used
MSA tool for performing multiple alignments. This program uses a progressive alignment system, in which it first compares sequences in pairs for significant similarities.
That identical pair of sequences is then viewed as a single sequence, and two-by-two
sequences are again compared together and matched in pairs. This process is repeated
until all sequences agree. With the default setting, a researcher may match the sequences, but as per the researcher's need, one can change the configuration to fit their
needs. The key parameters that can be modified are the matrix and penalties for gap
opening and gap extension. Clustal programs provide many input/output format options, including Clustal, PHYLIP (output only), and FASTA (input only). Using the FASTA
format for ClustalW input sequences, however, is best. Judging a sequence alignment's

consistency, however, is ultimately an informed guesswork. The bottom row of MSA ClustalW production includes stars (*), colons (:), and dots (.). A star below a column indicates a fully preserved or invariant amino acid residue, a colon (:) indicates that all residues in the column are roughly the same size and hydrophobicity, and a dot (.) indicates that the different amino acid residues in the column are either identical in size or hydrophobicity, while a lack of a symbol indicates that residues in the column vary in size and hydrophobicity. A basic sequence block criterion with good alignment is as follows: it is a gap-free continuous stretch of 10–30 amino acids with 1–3 stars, 5–7 colons, and a few dots scattered throughout the line. The matched sequences can also be viewed through the CLC Sequence Viewer 6.8. It uses color code to represent amino acids of similar size and hydrophobicity using the same color. Additionally, the conservation degree in each column is described as a bar diagram; completely preserved columns are represented by a maximum bar (100%) height. The Clustal Omega is a recent addition to the Clustal family. This high-capacity system is capable of aligning hundreds of thousands in a few hours with accuracy. Multiple processors are used for the alignment procedure, and better alignments are obtained, which are always reliable and superior to earlier models. Clustal Omega only has a command line interface and only manages protein sequences. Operating with protein sequences is superior to dealing with nucleotide sequences. Precompiled executables and software source codes for Windows, Linux, and Mac OS X systems are available from www.cluster.org. T-Coffee (igs-server.cnrs-mrs.fr/Tcoffee), MEME (http://meme.sdsc.edu/meme/website/intro.html; uses the approach of optimizing expectations), HMMER (http://hmmer.janelia.org/; for protein sequence analysis), MUSCLE (http://www.drive5.com/muscle/; http://www.ebi.ac.uk/tools/muscle/index.html; aligns all DNA and protein sequences), MAP (http://genome.cs.mtu.edu/map.html), and NCBI (for protein sequence alignments) are some of the other programs that can be used for MSA [73].

## 2.5 Conclusions

Due to the massive amount of data generated through modern genomic research, there is an urgent need to develop newer computational and algorithmic approaches to effectively manage and handle this data, which has been made possible through current advancements in the field of bioinformatics. With the advent of biological databases and data retrieval systems, it is possible to store a set of clearly defined data in an organized manner. The genome-sequencing projects have triggered high-throughput technologies that have produced sequence data at an unparalleled rapid pace, including the creation of computer programs capable of collecting, processing, classifying, and storing very large quantities of data, along with extracting information from the stored data. Moreover, modern bioinformatics tools need to be more advanced to deal with high-throughput metagenomics, proteomics, and metabolomics.

# References

[1]    Shaik N.A., et al. Introduction to Biological Databases. In Essentials of Bioinformatics, Volume I, Springer; 2019, 19–27.

[2]    Zou D., et al. Biological databases for human research. *Genomics, Proteomics & Bioinformatics*, 2015. **13**(1): 55–63.

[3]    Kodama Y., et al. The sequence read archive: explosive growth of sequencing data. *Nucleic Acids Research*, 2012. **40**(Database issue): D54–D56.

[4]    Pruitt K.D., et al. RefSeq: An update on mammalian reference sequences. *Nucleic Acids Research*, 2014. **42**(Database issue): D756–D763.

[5]    Knudsen M., Wiuf C. The CATH database. *Human Genomics*, 2010. **4**: 207–212.

[6]    Yang J.-H., et al. ChIPBase: A database for decoding the transcriptional regulation of long non-coding RNA and microRNA genes from ChIP-Seq data. *Nucleic Acids Research*, 2013. **41**(D1): D177–D187.

[7]    Sherry S.T., et al. dbSNP: The NCBI database of genetic variation. *Nucleic Acids Research*, 2001. **29**(1): 308–311.

[8]    Miyazaki S., et al. DNA Data Bank of Japan (DDBJ) in XML. *Nucleic Acids Research*, 2003. **31**(1): 13–16.

[9]    Vucetic S., et al. DisProt: A database of protein disorder. *Bioinformatics*, 2005. **21**(1): 137–140.

[10]   Karp P.D., et al. The EcoCyc database. *Nucleic Acids Research*, 2002. **30**(1): 56–58.

[11]   Kanz C., et al. The EMBL nucleotide sequence database. *Nucleic Acids Research*, 2005. **33**(suppl_1): D29–D33.

[12]   Zerbino D.R., et al. Ensembl 2018. *Nucleic Acids Research*, 2018. **46**(D1): D754–D761.

[13]   Consortium F. The FlyBase database of the Drosophila genome projects and community literature. *Nucleic Acids Research*, 2003. **31**(1): 172–175.

[14]   Benson D.A., et al. GenBank. *Nucleic Acids Research*, 2011. **39**(Database issue): D32.

[15]   Harrow J., et al. GENCODE: Producing a reference annotation for ENCODE. *Genome Biology*, 2006. **7**(1): S4.

[16]   Safran M., Dalah I., Alexander J., et al. GeneCards Version 3: The human gene integrator. *Database (Oxford)*, 2010. doi:. doi: 10.1093/database/baq020.

[17]   Camon E., et al. The gene ontology annotation (GOA) database: Sharing knowledge in Uniprot with Gene Ontology. *Nucleic Acids Research*, 2004. **32**(suppl_1): D262–D266.

[18]   Mulder N.J., et al. New developments in the InterPro database. *Nucleic Acids Research*, 2007. **35**(suppl_1): D224–D228.

[19]   Kanehisa M. The KEGG Database. In Novartis Foundation Symposium, Wiley Online Library, 2002.

[20]   Ogata H., et al. Computation with the KEGG pathway database. *Biosystems*, 1998. **47**(1–2): 119–128.

[21]   Kanehisa M., Goto S. KEGG: Kyoto encyclopedia of genes and genomes. *Nucleic Acids Research*, 2000. **28**(1): 27–30.

[22]   Gonzales M.D., et al. The Legume Information System (LIS): An integrated information resource for comparative legume biology. *Nucleic Acids Research*, 2005. **33**(suppl_1): D660–D665.

[23]   Karp P.D., et al. The MetaCyc database. *Nucleic Acids Research*, 2002. **30**(1): 59–61.

[24]   Pruitt K., Murphy T., Brown G., et al. RefSeq Frequently Asked Questions (FAQ), 2010 https://www.ncbi.nlm.nih.gov/books/NBK50679.

[25]   Liu C., et al. NONCODE: An integrated knowledge database of non-coding RNAs. *Nucleic Acids Research*, 2005. **33**(suppl_1): D112–D115.

[26]   Hamosh A., et al. Online Mendelian inheritance in man (OMIM). *Human Mutation*, 2000. **15**(1): 57–61.

[27]   Sussman J.L., et al. Protein Data Bank (PDB): Database of three-dimensional structural information of biological macromolecules. *Acta Crystallographica Section D: Biological Crystallography*, 1998. **54**(6): 1078–1084.

[28]   Bateman A., et al. The Pfam protein families database. *Nucleic Acids Research*, 2004. **32**(suppl_1):
       D138–D141.
[29]   Winnenburg R., et al. PHI-base: A new database for pathogen host interactions. *Nucleic Acids
       Research*, 2006. **34**(suppl_1): D459–D464.
[30]   Schaefer C.F., et al. PID: The pathway interaction database. *Nucleic Acids Research*, 2009. **37**(suppl_1):
       D674–D679.
[31]   Barker W.C., et al. The protein information resource (PIR). *Nucleic Acids Research*, 2000. **28**(1): 41–44.
[32]   McDowall M.D., et al. PomBase 2015: Updates to the fission yeast database. *Nucleic Acids Research*,
       2015. **43**(D1): D656–D661.
[33]   Attwood T.K., et al. PRINTS-S: The database formerly known as PRINTS. *Nucleic Acids Research*, 2000.
       **28**(1): 225–227.
[34]   Hulo N., et al. The PROSITE database. *Nucleic Acids Research*, 2006. **34**(suppl_1): D227–D230.
[35]   Geer L.Y., et al. The NCBI biosystems database. *Nucleic Acids Research*, 2010. **38**(suppl_1):
       D492–D496.
[36]   Roberts R.J., Vincze T., Posfai J., Macelis D. REBASE – a database for DNA restriction and
       modification: Enzymes, genes and genomes. *Nucleic Acids Research*, D234–6. doi:. doi: 10.1093/nar/
       gkp874.
[37]   Nawrocki E.P., et al. Rfam 12.0: Updates to the RNA families database. *Nucleic Acids Research*, 2015.
       **43**(D1): D130–D137.
[38]   Twigger S., et al. Rat Genome Database (RGD): Mapping disease onto the genome. *Nucleic Acids
       Research*, 2002. **30**(1): 125–128.
[39]   O'Leary N.A., et al. Reference sequence (RefSeq) database at NCBI: Current status, taxonomic
       expansion, and functional annotation. *Nucleic Acids Research*, 2016. **44**(D1): D733–D745.
[40]   Cherry J.M., et al. SGD: Saccharomyces genome database. *Nucleic Acids Research*, 1998. **26**(1): 73–79.
[41]   Lo Conte L., et al. SCOP: A structural classification of proteins database. *Nucleic Acids Research*, 2000.
       **28**(1): 257–259.
[42]   Consortium U. UniProt: A hub for protein information. *Nucleic Acids Research*, 2015. **43**(D1):
       D204–D212.
[43]   Lawson D., et al. VectorBase: A data resource for invertebrate vector genomics. *Nucleic Acids
       Research*, 2009. **37**(suppl_1): D583–D587.
[44]   Chen N., et al. WormBase: A comprehensive data resource for Caenorhabditis biology and
       genomics. *Nucleic Acids Research*, 2005. **33**(suppl_1): D383–D389.
[45]   Bowes J.B., et al. Xenbase: A Xenopus biology and genomics resource. *Nucleic Acids Research*, 2007.
       **36**(suppl_1): D761–D767.
[46]   Geer R.C., Sayers E.W. Entrez: Making use of its power. *Briefings in Bioinformatics*, 2003. **4**(2):
       179–184.
[47]   Etzold T., Ulyanov A., Argos P. SRS: Information retrieval system for molecular biology data banks.
       *Methods in Enzymology*, 1996. **266**: 114–128.
[48]   Fujibuchi W., et al., DBGET/LinkDB: An integrated database retrieval system. *Pacific Symposium on
       Biocomputing*, 1998: 683–694.
[49]   Romiti M., Cooper P. Search Field Descriptions for Sequence Database. 2010. https://www.ncbi.nlm.
       nih.gov/books/NBK49540.
[50]   Cochrane G., et al. Evidence standards in experimental and inferential INSDC Third Party Annotation
       data. *Omics: A Journal of Integrative Biology*, 2006. **10**(2): 105–113.
[51]   Benson D.A., et al. GenBank. *Nucleic Acids Research*, 2005. **33**(suppl_1): D34–D38.
[52]   Reisinger F., Martens L. Database on demand – an online tool for the custom generation of FASTA-
       formatted sequence databases. *Proteomics*, 2009. **9**(18): 4421–4424.
[53]   Mundy D.P., Chadwick D., Smith A., 2003. Comparing the performance of abstract syntax notation
       one (ASN. 1) vs eXtensible Markup Language (XML). *Proceedings Terena Networking Conference*.

[54] Garg V.K., et al. MFPPI–Multi FASTA ProtParam Interface. *Bioinformation*, 2016. **12**(2): 74.

[55] Gilbert D. Sequence file format conversion with command-line readseq. *Current Protocols in Bioinformatics*, **2003**(1): A. 1E. 1–A. 1E. 4.

[56] Wang L., Riethoven -J.-J., Robinson A. XEMBL: Distributing EMBL data in XML format. *Bioinformatics*, 2002. **18**(8): 1147–1148.

[57] Aiyar A. The Use of CLUSTAL W and CLUSTAL X for Multiple Sequence Alignment. In Bioinformatics Methods and Protocols, Springer; 2000, 221–241.

[58] Boutet E., et al. Uniprotkb/swiss-prot. In Plant Bioinformatics, Springer; 2007, 89–112.

[59] Bininda-Emonds O. *seqConverter. pl, version 1*. Institut fur Spezeille Zoologie und Evolutionsbiologie mit Phyletischem Museum. *Friedrich-Schiller-Universitat Jena*, 2006.

[60] Rice P., Longden I., Bleasby A. EMBOSS: The European molecular biology open software suite. *Trends in Genetics*, 2000 Jun 1, **16**(6): 276–277.

[61] Barker G., et al. Redundancy based detection of sequence polymorphisms in expressed sequence tag data using autoSNP. *Bioinformatics*, 2003. **19**(3): 421–422.

[62] Savage D., et al. SNPServer: A real-time SNP discovery tool. *Nucleic Acids Research*, 2005. **33**(suppl_2): W493–W495.

[63] Duran C., et al. AutoSNPdb: An annotated single nucleotide polymorphism database for crop plants. *Nucleic Acids Research*, 2009. **37**(suppl_1): D951–D953.

[64] Thiel T., et al. SNP2CAPS: A SNP and INDEL analysis tool for CAPS marker development. *Nucleic Acids Research*, 2004. **32**(1): e5–e5.

[65] Bradbury P.J., et al. TASSEL: Software for association mapping of complex traits in diverse samples. *Bioinformatics*, 2007. **23**(19): 2633–2635.

[66] Pritchard J.K., et al. Association mapping in structured populations. *The American Journal of Human Genetics*, 2000. **67**(1): 170–181.

[67] Saeed A., et al. TM4: A free, open-source system for microarray data management and analysis. *Biotechniques*, 2003. **34**(2): 374–378.

[68] Thimm O., et al. MAPMAN: A user-driven tool to display genomics data sets onto diagrams of metabolic pathways and other biological processes. *The Plant Journal*, 2004. **37**(6): 914–939.

[69] Burge C., Karlin S. Prediction of complete gene structures in human genomic DNA. *Journal of Molecular Biology*, 1997. **268**(1): 78–94.

[70] Chenna R., et al. Multiple sequence alignment with the Clustal series of programs. *Nucleic Acids Research*, 2003. **31**(13): 3497–3500.

[71] Edgar R.C., Batzoglou S. Multiple sequence alignment. *Current Opinion in Structural Biology*, 2006. **16**(3): 368–373.

[72] Thompson J.D., et al. The CLUSTAL_X windows interface: Flexible strategies for multiple sequence alignment aided by quality analysis tools. *Nucleic Acids Research*, 1997. **25**(24): 4876–4882.

[73] Edgar R.C. MUSCLE: A multiple sequence alignment method with reduced time and space complexity. *BMC Bioinformatics*, 2004. **5**(1): 113.

[74] Sharma V.R. Bioinformatics and its applications in environmental science and health and its applications in other disciplines. 2021. 88–93.

## Multiple choice questions

Q1  Which of the following databases is commonly used for storing and retrieving genetic sequence information?
   a)  PubMed
   b)  GenBank
   c)  UniProt
   d)  PDB

Q2  Which database contains information about known protein structures?
   a)  GenBank
   b)  UniProt
   c)  PDB
   d)  RefSeq

Q3  Which database is a comprehensive resource for protein sequence and annotation data?
   a)  GenBank
   b)  UniProt
   c)  PDB
   d)  BLAST

Q4  The NCBI hosts which of the following databases?
   a)  GenBank
   b)  UniProt
   c)  PDB
   d)  All of the above

Q5  Which database is commonly used for identifying similar sequences in a given dataset?
   a)  GenBank
   b)  UniProt
   c)  PDB
   d)  BLAST

Q6  Which database contains information about known genetic variations in human populations?
   a)  dbSNP
   b)  GenBank
   c)  UniProt
   d)  PDB

Q7  Which database provides information about the functions and interactions of proteins?
   a)  GenBank
   b)  UniProt
   c)  PDB
   d)  STRING

Q8  Which database contains information about known biological pathways and networks?
   a)  KEGG
   b)  GenBank
   c)  UniProt
   d)  PDB

Q9  Which of the following databases is commonly used for storing and analyzing genomic data of
    model organisms?
    a)  GenBank
    b)  Ensembl
    c)  UniProt
    d)  PDB

Q10  What is the primary function of the sequence retrieval system (SRS)?
    a)  To perform multiple sequence alignment
    b)  To predict protein structures
    c)  To retrieve and integrate biological data from various databases
    d)  To analyze gene expression patterns

**Answers**

Q1  b)  GenBank
Q2  c)  PDB
Q3  b)  UniProt
Q4  d)  All of the above
Q5  d)  BLAST
Q6  a)  dbSNP
Q7  d)  STRING
Q8  a)  KEGG
Q9  b)  Ensembl
Q10  c)  Integration of data from different sources into a unified platform

Varruchi Sharma*, Poonam Bansal, Imran Sheikh, Anupam Sharma,
Damanjeet Kaur, Amit Joshi, and Anil K. Sharma*

# Chapter 3
# Fundamentals of bioinformatics

**Abstract:** In a modern biological context, it is quintessential to have quantitative tools to analyze and compute genomic data. Various fields such as mathematics, statistics, biology, and computational tools are integrated to manage the enormous amount of high-throughput data quantitatively and in a timely fashion. This has further led to newer paths of research and development, especially in the OMICS field, answering specific queries related to biology, transforming the proteomics, genomics, and metabolomics fields, and bringing newer paradigms to human health. Bioinformatics is an interdisciplinary field of science as it conglomerates biology, computer science, information engineering, mathematics, and statistics for investigating and understanding biological data. Mathematical and statistical tools have been employed successfully to resolve biological queries through in silico analysis. Bioinformatics has shown a lot of promise in resolving many complex molecular biology queries through the merger and integration of mathematical and statistical tools. This chapter brings forth the fundamental aspects of bioinformatics, including various programs and methods associated with such programs and their implications in bioinformatics.

**Keywords:** Genomic data, quantitative tools, OMICS, in-silico analysis, fundamentals of bioinformatics

## 3.1 Introduction

In the modern era of biology, quantitative tools are of utmost importance. In other words, we can say that the understanding of biology has become extremely important in today's world. The integration of various fields such as mathematics, biology, and

*Corresponding author: Varruchi Sharma, Department of Biotechnology and Bioinformatics, Sri Guru Gobind Singh College, Sector 26, Chandigarh 160019, India, e-mail: sharma.varruchi@gmail.com
*Corresponding author: Anil K. Sharma, Department of Biotechnology, Amity University Punjab, Mohali 140306, Punjab, India, email: anibiotech18@gmail.com
Poonam Bansal, Department of Biotechnology, MMEC, Maharishi Markandeshwar (Deemed to be University), Mullana, Ambala, 133207, Haryana, India
Imran Sheikh, Department of Biotechnology, Eternal University, Baru Sahib, Sirmour, Himachal Pradesh, India
Anupam Sharma, Department of Physics, Guru Kashi University, Talwandi Sabo, Bathinda, Punjab, India
Damanjeet Kaur, Amit Joshi, Department of Biotechnology and Bioinformatics, Sri Guru Gobind Singh College, Sector 26, Chandigarh 160019, India

https://doi.org/10.1515/9783111568584-003

computer science has led to new paths of research and understanding. Moreover, these tools have helped in analyzing and processing information and answering specific queries related to biology. Bioinformatics is an interdisciplinary field of science integrating biology, computer science, information engineering, mathematics, and statistics for investigating and understanding biological data. The in silico analyses of biological queries are performed using mathematical and statistical techniques. The expansion of bioinformatics as a promising field is a result of the merger and integration of computer science and molecular biology employing specific mathematical and statistical tools. Therefore, bioinformatics is the study of biology using computational tools to retrieve information from biological data. It comprises the assembly, storage, retrieval, and data modeling for the investigation, imagining, or estimation through a particular procedure or a set of rules/instructions in the form of algorithms or software. The significant areas in bioinformatics aim to comprehend complex biological systems and mechanisms. However, the use of computational tools with respect to biology does not conclude the similarity of bioinformatics with computational biology. Bioinformatics concludes the data at sequential, structural, and functional levels of genes and genomes. However, the other field, computational biology, focuses on the computation of biological data, and this also limits the hypothetical expansion of algorithms in bioinformatics [1].

The goal of bioinformatics is to understand biological data (biological processes) and make that data suitable for biological databases, database creation, and progression, followed by the advancement of computational methods.

## 3.2 Bioinformatics association with other domains

Bioinformatics is synonymous with many fields, as it is an integration of different disciplines. Having similarities with computational biology, the computation of biological data is carried out for the purpose of bioengineering in order to access and better understand biological data on computers. The biological data analysis in computational biology comprises sequential information/data. The sequential information in use is of DNA, RNA, and the sequences of proteins. Such analysis of data or the extraction of significant information has been made possible by various programs and the use of algorithms, data mining approaches, software, and so on. Bioinformatics has also helped in the management and investigation of genomic data, which includes the annotations at the genome level, and expression profiling. This analysis has been performed using the expansion of further resourceful and more promising bioinformatics approaches. Without the help of bioinformatics tools, it could not have become possible to handle large-scale data either associated with sequencing or other sort of analysis. This field has not only delivered a well-defined tool to work with, investigate, equate, and find associations among data and visualizations of sequential data but also processes the progression of sequencing of data. In the present scenario, the next-generation sequencing has facilitated the decryption of the whole genome among diverse organisms. Tools in bioinformatics also help in analyzing gene and protein expression profiles. Moreover, bioinformatics modeling at the structural level along with designing of molecules has also been performed (Table 3.1).

**Table 3.1:** Frequently used programs and associated methodologies in the field of bioinformatics.

| S. no. | Process | Program used | Methodology |
|---|---|---|---|
| 1 | Genome sequencing | BLAST | Alignment |
| | | Next-generation sequencing (NGS) | Decryption of the whole genome among diverse organisms |
| | | Genomes OnLine Database (GOLD) | Genome decode |
| 2 | Analysis of gene and protein expression profiles | Serial analysis of gene expression (SAGE) | Characterization of functional elements of the human genome |
| | | Expressed sequence tags (ESTs) | |
| | | Transcriptome profiling | |
| | | Massively parallel signature sequencing (MPSS) | |
| | | Encyclopedia of DNA elements (ENCODE) | |
| 3 | Homology modeling | SWISS-MODEL | Automated webserver (based on ProModII) |
| | | ROBETTA | |
| | | MODELLER | Program for protein modeling; mainly written in Fortran and Python |
| | | FoldX | Energy calculations and protein design |
| 4 | Threading/fold recognition | Iterative Threading ASSEmbly Refinement (I-TASSER) | Structural and function predictions |
| 5 | Secondary structure prediction methods | Hidden Markov model | Ab initio based and homology based |
| | | Neural network method | |
| | | Support vector machines | |
| 6 | Signal peptide prediction | Artificial neural networks and hidden Markov models | Predicts the presence of signal peptides and the locations of their cleavage sites in proteins |

Bioinformatics software and tools that are user-friendly and capable of performing extended and integrated analyses with better visualization and graphical outputs

**Table 3.1** (continued)

| S. no. | Process | Program used | Methodology |
|--------|---------|--------------|-------------|
| | | **Open-source software** | |
| | UGENE | UNIGENE is a free, open-source software for DNA/protein sequence, alignment, assembly, visualization, and annotation | |
| | EMBOSS | This is a free, open-source software for sequence analysis | |
| | GenGIS | With the help of this program, users can merge digital map data with environmental data regarding biological sequences | |
| | GENtle | It is a software for DNA and amino acid editing, database management, plasmid maps, restriction and ligation, alignments, sequencer data import, calculators, gel image display, and PCR | |
| | MOTHUR | It calculates the probable misaligned matches in the sequences | |
| | BioPerl | A collection of Perl modules called BioPerl makes it easier to create Perl scripts for bioinformatics applications | |
| | PathVisio | Tool for pathway analysis and visualization | |
| | BioJava | The common bioinformatics operations of converting DNA sequences into proteins, reading and writing popular sequence file formats, and more are all supported by BioJava | |
| | GenoCAD | Based on the principle of formal languages, GenoCAD makes it easier to build protein expression vectors, synthetic gene networks, and other genetic constructions for genetic engineering | |
| | Biopython | The biggest and most used bioinformatics library for Python is called Biopython. For typical bioinformatics activities, it contains a variety of submodules. It was primarily written in Python and was created by Chapman and Chang. Additionally, C code is included to enhance the software's sophisticated computing section | |
| | geWorkbench | A Java-based open-source platform called geWorkbench (genomics Workbench) allows plug-ins to be customized into sophisticated bioinformatics applications | |
| | GenomeSpace | An intuitive web interface is provided by GenomeSpace, a cloud-based interoperability system that supports integrative genomics analysis | |
| | Bioclipse | This is a Java-based project, an open-source visual platform for chemo- and bioinformatics (proteins, sequential data, molecular data, spectra, and scripts) based on the Eclipse Rich Client Platform | |
| | .NET Bio | It is a library that was created to enable simple loading and analyzing the biological data | |
| | Apache Taverna | It is an open-source software tool that is used for designing workflows and also for their execution. It was created by the myGrid project | |

**Table 3.1** (continued)

| S. no. | Process | Program used | Methodology |
|---|---|---|---|
| | BioJS | | It is an open-source project (maintained by small building blocks) for bioinformatics data on the web |
| | Bioconductor | | It is a free, open-source software project that helps in analyzing the genomic data generated by wet lab experiments in molecular biology |
| | BioRuby | | It is an open-source Ruby code with a set of free development tools along with libraries for bioinformatics and molecular biology data |
| | | **Sources used for sequence alignment** | |
| 1 | Multiple sequence alignment (MSA) | | Alignment of three or more biological protein/DNA sequences of similar length. There are two main approaches to MSA, which cover progressive and iterative MSAs |
| 2 | Sequence search services (SSS) | | Demonstrate and apply web-based bioinformatics solutions |
| 3 | Bioinformatics workflow management systems (BWMS) | | Designed for the construction and accomplishment of data operation |
| 4 | Biological networks | | A network that applies to biological systems |
| 5 | System biology | | Analysis and modeling of complex biological systems at a computational and mathematical level |
| 6 | Text mining | | Using programs designed in bioinformatics to assemble and constitute the biomedical literature that helps in understanding and inquiring about the data of interest |
| 7 | Bioinformatics Training Network (BTN) | | Provides bioinformatics training (essential educational materials) |
| 8 | The Global Organization for Bioinformatics Learning, Education, and Training (GOBLET) | | Educational spread globally |

# 3.3 Applications of bioinformatics

Bioinformatics is a multidisciplinary field that combines biology, computer science, mathematics, statistics, and so on. This integration plays a noteworthy role in data analysis, interpretation, and management at a biological level. By this time, the field of bioinformatics has transformed various characteristics of life sciences. Additionally, it continues to make a substantial contribution to other fields of science. The key application of bioinformatics is the examination of the sequential characteristics that include the DNA and protein sequences of an organism. The comparison or alignment of sequences assists as a foundation for a number of manipulations while comparing the sequences [1]. Furthermore, the alignments of sequences of either DNA or protein help in identifying various features or resolving diverse problems, such as structural and functional character determination of novel sequences, and finding out the best evolutionary relationships [2, 3].

Some prominent applications of the field of bioinformatics are as follows:

Genomics, proteomics, and transcriptomics are the most well-established applications of bioinformatics. Various tools are used to analyze and interpret DNA sequences, and identify genes, regulatory elements, genetic variations, and so on. The proteomic data involves the large-scale analysis of proteins expressed by an organism or cell [4]. Gene, genome, protein structure, and function prediction assist in drug discovery and understanding protein–protein interactions. This information is instrumental in understanding the genetic basis of diseases, evolution, population genetics, and so on. In transcriptomics, the RNA transcripts of an organism are of major concern. In bioinformatics, various tools are used for analyzing sequential data at the RNA level. The study of gene expression patterns under different conditions is also a major application of the field [5].

Drug development and drug designing: In this field, the drug design and discovery process is performed by facilitating virtual screening of the best probable drug candidates. The process is carried out in both ways, such as receptor-based and ligand-based drug design [6, 7].

Phylogenetics: In the field of bioinformatics, the evolutionary relationships among various organisms/species can be studied. With the help of molecular data, we can understand the evolutionary history of an organism. Evolutionary history helps in determining species relationships, and this also leads to outlining the major cause (root) of diseases.

Structural biology: It involves predicting and understanding the three-dimensional (3D) structures of proteins and nucleic acids. The structure-level understanding of DNA/protein is vital for drug design and the rational drug discovery process. Once we have a deep understanding of genomic and genetic alteration levels, with the help of bioinformatics, we can tailor treatments based on an individual's genetic makeup, also called Personalized Medicine [8].

Metagenomics: It involves understanding the genetic content of whole microbial communities. It is a major application in environmental studies. It helps in gaining insights into the human microbiome and investigating the diversity of microorganisms [9].

Systems biology: It is the study of complex biological systems, investigating interactions among genes, proteins, and other molecules. These interactions help in understanding how they function together to perform various biological processes [10].

# 3.4 Database development

In bioinformatics, database construction plays a significant role in data storage, management, and analysis of biological and genomic data. The data generated from various high-throughput technologies is difficult to manage, design, and implement for the scientific community in order to access and interpret biological information effectively (Figure 3.1).

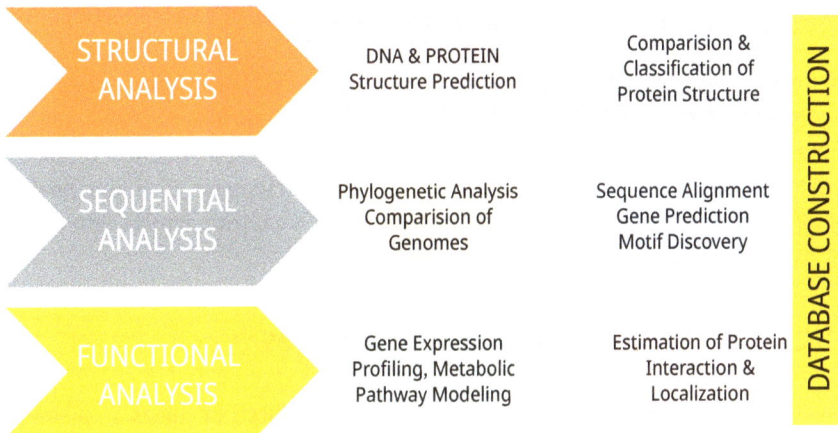

**Figure 3.1:** Database construction.

Database development in bioinformatics involves the following:

Data integration: In bioinformatics, it is quite difficult to integrate the data that we get from various sources. In database construction, the data from different experiments, laboratories, and studies are merged in order to create a unified resource for research purposes [11].

Data diversity: In the databases, a vast collection of biological data consisting of DNA sequences, protein structural data, gene expression data, genetic variations, and so on is included. For each data type, specific data models and storage mechanisms are required to accommodate the different biological information [12].

Annotation and consistency: In the databases, there should be community-accepted standards to ensure data consistency. Additionally, comprehensive annotation of biological entities and metadata is essential to provide context and relevance to the stored information [13].

Query and retrieval: In database construction, designing efficient querying systems is a matter of deep concern in order to retrieve relevant information quickly. Database optimization is most important for accuracy and speed in the database [14].

Data security and privacy: Keeping in view the sensitivity of data (biological) and its best probable application in healthcare and research, the database must incorporate robust security measures. The security methods opted by the database act as a safe-guard for information from unlawful access and maintain privacy. Secure data also improves at the performance level. As the data grows in the database, the construction of the database should anticipate scalability requirements. The ever-increasing data size in research, while maintaining optimal performance, is essential for long-term usability in constructing databases. The databases constructed must have web accessibility in order to make them available to the scientific community through a web interface [15].

The biological database construction is an ongoing process that integrates bioinformatics, biological data, and database experts. By considering the specific challenges of data management, databases have become a helpful resource in scientific research. These databases significantly contribute to advancing diverse fields, including healthcare and management systems, medicine, agriculture, and environmental sciences.

## 3.5 Sequence alignment and its types

Alignment is an agreement or a comparison between two objects or things. In bioinformatics, "sequence alignment" is the arrangement of sequences of either DNA/RNA or protein, which can help in identifying the regions of similarity/likeness. These similar regions can correspond to structural, functional, or evolutionary associations among sequences.

In this process, the sequences are equated by examining those character patterns that are common and express residue-by-residue correspondence within the related sequences. The alignments can be performed between two reference sequences (known sequences), between two unknown sequences (query sequences), or between a reference and a query sequence [16].

The following are the three main types of alignment:

Pairwise sequence alignment: It involves comparing two similar sequences to determine their degree of similarity and identify conserved regions at a particular time. There are two main algorithms of pairwise alignment that work at the backend for performing alignments: (a) Needleman–Wunsch algorithm, which is used for global alignments. In global alignment, the entire length of both sequences is considered for finding the best alignment over the entire length of the sequence. (b) Smith–Waterman algorithm, which is used for the execution of local alignments. In local alignment, a specific region or a particular segment of the sequence is taken into consideration. This is significant in alignments when the conserved regions are surrounded by less similar or dissimilar regions (Figure 3.2) [17].

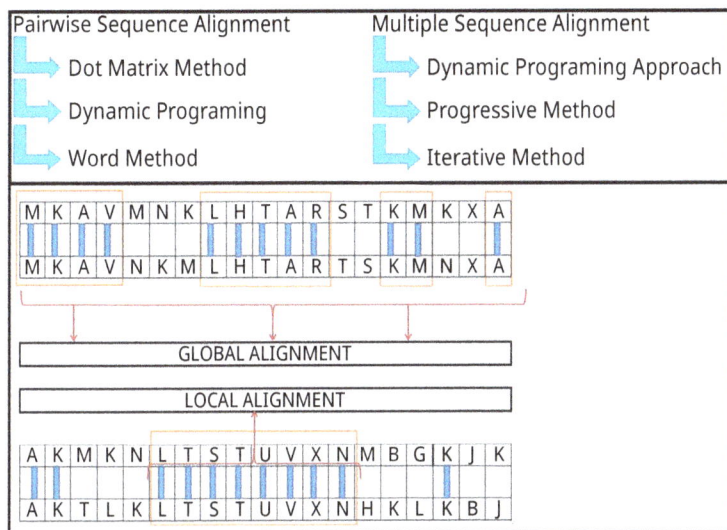

**Figure 3.2:** Demonstration of the alignment procedure.

Multiple sequence alignment (MSA): MSA involves the alignment of three or more sequences at a time. The MSA procedure is significant in studying evolutionary relationships among various sequences, recognizing conserved regions, and predicting the structure and function of proteins. There are mainly four algorithms that majorly help in performing sequence alignments: ClustalW, MUSCLE, T-Coffee, and MAFFT [18].

Progressive alignment: This method of pairwise sequence alignment is a heuristic approach that is used to construct MSA. In this method, first the alignment of the most similar sequences is considered, and then the remaining sequences are added step by step in the alignment procedure. The progressive method of alignment, or the hierarchical method, is more efficient for larger sets of sequences. Sequence alignment tools help in understanding the evolutionary history of sequences, identifying conserved regions in the protein or DNA sequences, and predicting protein structure [19].

## 3.6 Scoring a sequence alignment

Scoring methods are an essential part of sequence alignment algorithms. These methods help in determining the quality of alignments among DNA, RNA, or protein sequences. Scoring methods aim to measure the degree of similarity or dissimilarity among DNA/protein sequences. After the completion of alignment, the assignment of scores to matches, mismatches, and gaps is performed using scoring methods. The scores generated play a critical role in identifying the optimal alignment [20].

There are numerous scoring methods that are used in sequence alignment (Figure 3.3).

Match and mismatch scores: In pairwise sequence alignment, a match score is assigned when two nucleotides or amino acid residues are identical, representing a conserved region. A mismatch score is assigned when the residues aligned are not the same, which represents a distinction among the sequences.

Gap penalty: In alignment, it is fairly common to introduce gaps, and these gaps help to account for insertions or deletions in DNA/protein sequences. Gap penalties are negative scores assigned for presenting gaps in the sequences.

Gap extension/opening penalty: The gap extension penalty, which can be less severe than the gap opening penalty, is a distinct score for extending an existing gap. It discourages the development of protracted gaps with no breaks. When a gap is introduced into an alignment for the first time, a penalty score is given. It affects the possibility of creating a fresh gap while aligning.

Linear/affine gap penalty: In the linear gap penalty method, a constant score (e.g., –1 or –2) is deducted for each gap in the alignment procedure. In the affine gap penalty, two scores are used: (i) a gap opening penalty and (ii) a gap extension penalty.

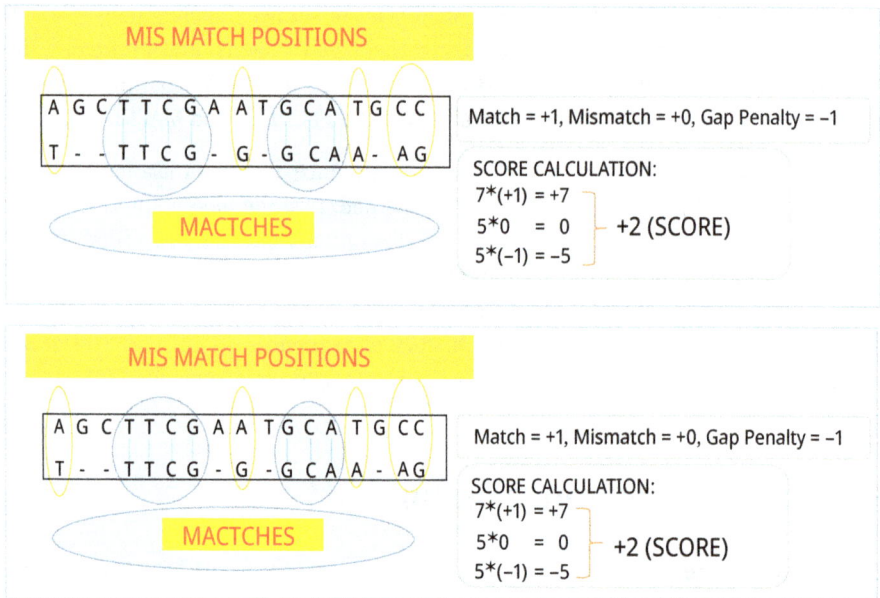

**Figure 3.3:** Alignment procedure.

Scoring matrices (substitution-based alignment methods): On the basis of observed frequencies in protein alignments, these matrices assign scores to all possible amino acid substitutions. BLOSUM (BLOcks SUbstitution Matrix) series and PAM are the most commonly used scoring matrices [19].

Scoring schemes for nucleotide alignments: Some other factors may be taken into account while aligning nucleotide sequences. In coding regions (exons), a greater score is assigned for matches than that of matches obtained in noncoding regions (introns).

There are a few important terms that are used in the alignment process (Table 3.2).

**Table 3.2:** Terms used in the alignment process.

| S. no. | Term used | Description | References |
|---|---|---|---|
| 1 | Match | Identification of the regions that are similar between the sequences | [21–29] |
| 2 | Mismatch | Identification of the regions that are not similar between the sequences | |
| 3 | Gap | Gap is the space that is put in an alignment to compensate for insertions and deletions. A letter, if paired with a null value, is called a gap (scores or the gap scores are usually negative) | |
| 4 | Score of alignment | It is the sum of substitution (PAM and BLOSUM methods) | |
| 5 | Identity | Extent of similarity of residue at the same position, expressed as a percentage | |
| 6 | Similarity | Relative similarity of a sequence (protein/nucleotide) is expressed as the percent identity of a sequence | |
| 7 | Substitution matrix | Rate at which one character in a sequence deviates to other character states over time | |
| 8 | Conservation | Changes at a specific position of an amino acid or (less commonly, DNA) sequence that preserve the physicochemical properties of the original residue | |
| 9 | Homology | Similarity attributed to descent from a common ancestor | |
| 10 | Orthologs | Homologous sequences in different species arose from a common ancestral gene during speciation and may or may not be responsible for a similar function | |

**Table 3.2** (continued)

| S. no. | Term used | Description | References |
|---|---|---|---|
| 11 | Paralogs | Homologous sequences within a single species that arose by gene duplication | |
| 12 | Hamming distances | Distances between two equivalent-length sequence threads; can also be defined as the number of spots with mismatching sequence characters | |
| 13 | Edit distances | Edit operations (deletion/insertion/alteration) are performed to change one sequence thread into another (threads may be of inadequate lengths/surely not of equal lengths) | |

Alignments are performed in order to find homologs. The searches in pairwise alignments deliver precise evaluations of the compared sequences.

The algorithms used for aligning both at the global and the local levels of alignments are basically similar, and the difference in the alignment lies in the optimization strategy. Both algorithms can be based on the dot matrix method, the dynamic programming method, and/or the word method (used for performing fast database similarity searching) [30].

For better/accurate results/alignments, more accurate tools are needed to evaluate the sequences in detail, which include (a) dot plots for the graphical analysis of data and (b) residue-by-residue exploration of sequences by local or global alignments [31].

PAM (point accepted mutation) and BLOSUM are two scoring matrices that are substitution-based alignment algorithms and are widely used for protein sequence alignments.

PAM scoring matrices were developed by Margaret Dayhoff in 1970. The point mutations in the protein sequences are the main focus of its construction. The method is probability-based, quantifying the possibility of a particular amino acid substitution occurring at a specific position in the protein sequence. PAM0, PAM1, PAM30, PAM80, PAM110, PAM200, PAM250, etc., are the most common versions of PAM matrices. In the PAM versions, the numbers assigned indicate the evolutionary distance. For example, PAM0 shows 0% mutational rate and 100% sequence identity, PAM1 shows 1% mutational rate and 99% sequence identity, and PAM250 shows 80% mutational rate and 20% sequence identity [32] (Table 3.3).

**Table 3.3:** PAM numbers with mutational rates and sequence identities.

| PAM number | 0 | 1 | 30 | 80 | 110 | 200 | 250 |
|---|---|---|---|---|---|---|---|
| Mutational rate | 0% | 1% | 25% | 50% | 40% | 75% | 80% |
| Sequence identity | 100% | 99% | 75% | 50% | 60% | 25% | 20% |

Higher PAM scores show more conservative and highly similar substitutions; on the other hand, lower scores indicate less similar and more divergent substitutions.

BLOSUM matrices were developed by Steven Henikoff and Jorja Henikoff in 1990. Unlike the PAM matrices method, BLOSUM matrices result from the analysis of local sequence alignments, or "blocks," from different protein sequences. This is a series of blocks of amino acid substitution matrices (BLOSUM), all derived based on direct observation. The observations were derived for every probable amino acid substitution in MSAs. The alignments constructed in the matrix method are based on conserved amino acid patterns, showing a diverse group of protein sequences. The patterns in the sequence are called blocks, which are identified as the ungapped alignments of less than 60 amino acid residues in length. BLOSUM62 is the most frequently used matrix for protein alignments, in which the value 62 represents the 62% identity of the sequence [33].

# 3.7 Dot plot method

A dot plot in the alignment is the graphical representation of a pairwise comparison of two sequences. The method reveals complex patterns among sequences. This is also a perfect method for checking out the features that may come across in altered orders. The sequences are compared in pairs, or two sequences at a time are compared to find the best matches among sequences. A dot plot delivers a good way to understand the alignment, but on the other hand, it does not arrange for residue-by-residue analysis [34].

The similarity in two sequences is evaluated in the following ways: (i) the number of nucleotides equated each time can also be defined as window size, which is frequently an odd number; (ii) for placing a dot on a matched position, it should be noted that a minimum number of nucleotides coordinated is defined as stringency; (iii) to check for mismatched positions, those nucleotides which are not matched, a dot is placed on such positions as well. Therefore, we can check for a mismatch limit that shows how much the window size is out of the stringency value [35].

In the simplest form, a dot is placed at the point where horizontal and vertical sequence values match. In cases where there are two sequences that have substantial regions of similarity, many dots line up, forming contiguous diagonal lines. These diagonal lines reveal the sequence alignment. Interruptions in the diagonal lines indicate insertions or deletions. On the other hand, insertions or deletions in the sequence are shown by parallel diagonal lines. This method is a sensitive qualitative indicator of similarity; rearrangements of sequences can be done. During the alignment, when we are comparing large sequences, a problem exists known as the high noise level (most common in DNA sequences due to its four bases). In the dot plot method, due to noise, the dots are plotted all over the graph, which confuses the identification of the true alignments.

In DNA sequences, the problem is acute because of the four (AGCT bases) characters. Due to only these four characters in the DNA sequence, there is a one in four chance of matching a particular residue in the sequence. The "noise" level reduction is performed using a filtering technique, in which a "window" of a particular length covers a section of residue pairs. A "window" is also known as a tuple [36].

In the filtering technique, the window of particular residues slides across the two sequences in order to compare all the best probable stretches of the sequences. In the alignment, the dots are placed at the time when a fixed stretch of residues equal to the window size from one sequence matches completely with a fixed length of another sequence. There are many software programs that help in creating dot plots (Figure 3.4).

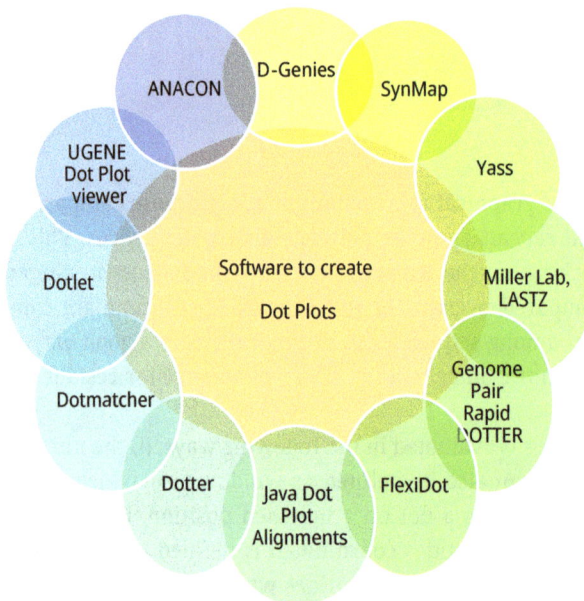

**Figure 3.4:** Software that helps in creating dot plots.

## 3.8 Applications of pairwise sequence alignment method

The method is used to compare and analyze two sequences at a time, identifying similarities, differences, and evolutionary relationships. The pairwise sequence alignment method has various applications such as phylogenetic analysis (to understand the evolutionary history among various species), comparative genomics (to identify regulatory regions, conserved genes, and other genomic elements, helping in understanding genome evolution and functions in organisms), protein structure/functional prediction (alignment with known structures helps in finding the 3D structure of the unknown protein, and models for unknown proteins are generated using homology modeling methods), functional domain identification (by aligning unknown protein sequences with proteins with known functions, we get conserved regions that correspond to specific functional domains in already known proteins), mutation analysis, functional annotation (to predict the function of the query sequence by alignment), and identification of homologous sequences (for inferring evolutionary relationships) [37, 38].

## 3.9 Dynamic programming

In dynamic programming, the focus is to solve a complicated problem by simplifying it into small components or making it a subproblem. Hence, this method is used for solving optimization problems and works on the principle of optimality. The same was proposed by Needleman and Wunsch for the first time, which pronounces the use of an algorithm for sequence alignment. The score calculation depicts the large number of nucleotides matching with others, displaying the maximal match. This method also constructs a two-dimensional (2D) alignment grid for calculating alignments. This method transfigures a dot matrix into a scoring matrix to account for matches and mismatches among sequences [39].

   This method of programming has been used for the alignment of sequences, protein folding, and structure prediction (RNA), along with the binding of protein and DNA. Moreover, this method constructs a 2D matrix, and the axes of the matrix are the two sequences that are to be compared. The calculation of scores is performed row-wise, which works upon the strategy of scanning one sequence from row one to the other sequence of another row. In the calculation of the score, sequence from the row one scans over the entire length of the sequence in the other row, and matches found in the alignment are calculated. The best score is recorded, and the procedure is repeated [40].

## 3.10 Longest common subsequence

If one needs to understand the similarity of a novel sequence in reference to a known one, longest common subsequence is the method of choice. A substring is defined first, in which the characters need to be contiguous, followed by a subsequence in which the characters do not need to be contiguous [41].

Iterative alignment: This alignment creates a progressive alignment and calculates a quality score after performing the alignment. In this alignment, one or more new sequences are added to the algorithm, and initially aligned sequences are repeatedly realigned so that the best alignment can be obtained. This method can correct alignment errors dynamically [42].

Some of the major characteristics of iterative alignment are discussed next.

For iterative alignment, the fast and approximate method (pairwise/progressive alignment) is used (known as seed alignment). In the iteration process, a profile or position-specific scoring matrix is constructed from pairwise/progressive alignment (seed alignment). The conserved positions are represented by profiles. Also, the propensity of observing different residues at every possible position is observed by the profiles. These profiles are searched over sequence databases to obtain homologous sequences. Furthermore, these newly identified sequences are added to seed alignment, where more new alignments are created using the profile-based alignment method. Multiple iterations are obtained using processes like profile construction, searching over the database, and alignment extension. As the iteration process moves on, the alignment gets refined, and the same process goes on until an already defined stopping criterion is met [43].

Progressive alignment: This is the most common method used for MSA. In this method, the most similar sequences are aligned together, and then related sequences are added one by one progressively. The alignment scores create a similarity matrix, which is used to produce a guide tree.

This alignment method uses a heuristic approach for performing alignments in which the most closely related sequences are aligned first, and then distantly related sequences are taken into consideration. The algorithm of progressive alignment is considered the most efficient and suitable set of instructions for aligning large sets of sequences [44].

Progressive alignment follows these steps for performing alignment:

Pairwise alignment: The start of progressive alignment begins with the execution of pairwise sequence alignment. The same process is accomplished using the Needleman–Wunsch or Smith–Waterman algorithm.

Distance calculation: After the completion of the first step of alignment, distance scores are calculated. These scores help us in determining the evolutionary distance or dissimilarities between each sequence and the reference sequence [45].

Guide tree construction: The hierarchical clustering methods construct a guide tree.

Progressive alignment: For the best alignment, gaps are introduced in each new sequence to ensure its alignment consistency with the others.

Scoring and optimization: Scoring/substitution matrices are used for score calculation.

Progressive alignment can handle a number of sequences at a time. With this, we can construct MSAs. The alignments generated have sequences with high similarities. Moreover, there are some limitations to the method, especially when dealing with highly divergent or distantly separated sequences [46].

## 3.11 Conclusions

Quantitative tools have been the mainstay for analyzing and computing high-throughput genomic data. What we require is the integration of mathematical and statistical tools, biology, and computational tools to manage the enormous amount of high-throughput data quantitatively and in a time-bound manner. This has further led to newer paths of research and development, especially in the OMICS field, answering specific queries related to biology and transforming the proteomics, genomics, and metabolomics fields, bringing newer paradigms to human health. This fascinating field of computational biology has been able to bring together computer science, information technology, mathematics, and statistics for investigating and understanding complex biological data. Mathematical and statistical tools have been employed successfully to resolve biological queries through in silico analysis, thus showing a lot of promise in resolving many complex molecular biology queries through the merger and integration of mathematical and statistical tools.

## References

[1]    Nisbet R., Elder J., Miner G. Handbook of Statistical Analysis and Data Mining Applications, Academic Press; 2009.
[2]    Lesk A. Introduction to Bioinformatics, Oxford University Press; 2019.
[3]    Gu J., Bourne P.E. Structural Bioinformatics, Vol. 44, John Wiley & Sons; 2009.
[4]    Sharma V., Sharma A.K., Yadav M., Sehrawat N., Kumar V., Kumar S., Gupta A., Sharma P., Chakrabarti S. Prediction models based on miRNA-disease relationship: Diagnostic relevance to multiple diseases including COVID-19. *Current Pharmaceutical Biotechnology*, 2023. **24**(10): 1213–1227.
[5]    Liu L., Tang L., Dong W., Yao S., Zhou W. An overview of topic modeling and its current applications in bioinformatics. *SpringerPlus*, 2016. **5**(1): 1–22.
[6]    Sharma V. PI3Kinase/AKT/mTOR pathway in breast cancer; pathogenesis and prevention with mTOR inhibitors. *Proceedings of IVSRTLSB-2021*, 2022. **7**(1): 184–191.
[7]    Sharma V., Panwar A., Sankhyan A., Ram G., Sharma A.K. Exploring the potential of chromones as inhibitors of novel coronavirus infection based on molecular docking and molecular dynamics simulation studies. *Biointerface Research in Applied Chemistry*, 2022. **13**(2): 1–8.
[8]    Che D., Liu Q., Rasheed K., Tao X. Decision tree and ensemble learning algorithms with their applications in bioinformatics. *Advances in Experimental Medicine and Biology*, 2011. **696**: 191–199. doi: 10.1007/978-1-4419-7046-6_19.

[9]    Aguiar-Pulido V., Huang W., Suarez-Ulloa V., Cickovski T., Mathee K., Narasimhan G. Metagenomics, metatranscriptomics, and metabolomics approaches for microbiome analysis. *Evolutionary Bioinformatics Online*, 2016. **12**(Suppl 1): 5–16. Published 2016 May 12. doi: 10.4137/EBO.S36436.

[10]   Jünemann S., Kleinbölting N., Jaenicke S., Henke C., Hassa J., Nelkner J., Stolze Y., Albaum S.P., Schlüter A., Goesmann A. Bioinformatics for NGS-based metagenomics and the application to biogas research. *Journal of Biotechnology*, 2017. **261**: 10–23.

[11]   Baxevanis A.D., Bader G.D., Wishart D.S. Bioinformatics. John Wiley & Sons; 2020.

[12]   Teufel A. Bioinformatics and database resources in hepatology. *Journal of Hepatology*, 2015. **62**(3): 712–719.

[13]   Rhee S.Y. Bioinformatics. Current limitations and insights for the future. *Plant Physiology*, 2005. **138**(2): 569–570.

[14]   Abdurakhmonov I.Y. Bioinformatics: Basics, Development, and Future, InTech Rijeka; 2016.

[15]   Ouzounis C.A., Valencia A. Early bioinformatics: The birth of a discipline – A personal view. *Bioinformatics*, 2003. **19**(17): 2176–2190.

[16]   Potla P., Ali S.A., Kapoor M. A bioinformatics approach to microRNA-sequencing analysis. *Osteoarthritis and Cartilage Open*, 2021. **3**(1): 100131.

[17]   Lakshmi P., Ramyachitra D. Review about Bioinformatics, Databases, Sequence Alignment, Docking, and Drug Discovery. 2020, doi: 10.1007/978-981-15-2445-5_2.

[18]   Daugelaite J., Driscoll O., Sleator A., R.d. An overview of multiple sequence alignments and cloud computing in bioinformatics. *ISRN Biomathematics*, 2013. 2013: 615630. doi: 10.1155/2013/615630.

[19]   Bonny T., Zidan M.A., Salama K.N. An Adaptive Hybrid Multiprocessor Technique for Bioinformatics Sequence Alignment, IEEE; 2010; 112–115.

[20]   Sharma K.R. Bioinformatics: Sequence Alignment and Markov Models, McGraw-Hill Education; 2009.

[21]   Fassler J., Cooper P. *BLAST Glossary. BLAST® Help; 20*11.

[22]   Korf I., Yandell M., Bedell J. Blast. "O'Reilly Media, Inc.", 2003.

[23]   Rustan A. Rock Blasting Terms and Symbols: A Dictionary of Symbols and Terms in Rock Blasting and Related Areas like Drilling, Mining and Rock Mechanics, CRC Press; 1998.

[24]   Mount D.W. Using BLOSUM in sequence alignments. *Cold Spring Harbor Protocols*, 2008. **2008**(6): pdb-top39.

[25]   Gogarten J.P., Olendzenski L. Orthologs, paralogs and genome comparisons. *Current Opinion in Genetics & Development*, 1999. **9**(6): 630–636.

[26]   Saigo H., Vert J.-P., Ueda N., Akutsu T. Protein homology detection using string alignment kernels. *Bioinformatics (Oxford, England)*, 2004. **20**(11): 1682–1689.

[27]   Li H. Aligning sequence reads, clone sequences and assembly contigs with BWA-MEM. arXiv preprint arXiv:1303.3997. 2013 Mar 16.

[28]   Jain C., Zhang H., Gao Y., Aluru S. On the complexity of sequence-to-graph alignment. *Journal of Computational Biology*, 2020 Apr 1. **27**(4): 640–654.

[29]   Giegerich R., Wheeler D. Pairwise sequence alignment. *BioComputing Hypertext Coursebook*, 1996. **2**: 1–6.

[30]   Autenrieth F., Isralewitz B., Luthey-Schulten Z., Sethi A., Pogorelov T. Bioinformatics and Sequence Alignment, University of Illinois at Urbana-Champaign; 2005, 1–29.

[31]   Pereira R., Oliveira J., Sousa M. Bioinformatics and computational tools for next-generation sequencing analysis in clinical genetics. *Journal of Clinical Medicine*, 2020. **9**(1): 132.

[32]   Mount D.W. Comparison of the PAM and BLOSUM amino acid substitution matrices. *Cold Spring Harbor Protocols*, 2008. **2008**(6): pdb. ip59.

[33]   Krane D.E., Raymer M.L. Protein Alignment Scoring-PAM and BLOSUM.

[34]   Sperschneider J., Datta A. DotKnot: Pseudoknot prediction using the probability dot plot under a refined energy model. *Nucleic Acids Research*, 2010. **38**(7): e103–e103.

[35] Leonard S., Lardenois A., Tarte K., Rolland A.D., Chalmel F. FlexDotPlot: A universal and modular dot plot visualization tool for complex multifaceted data. *Bioinformatics Advances*, 2022. **2**(1): vbac019.

[36] Ivry T., Michal S., Avihoo A., Sapiro G., Barash D. An image processing approach to computing distances between RNA secondary structures dot plots. *Algorithms for Molecular Biology*, 2009. **4**: 1–19.

[37] Chatzou M., Magis C., Chang J.-M., Kemena C., Bussotti G., Erb I., Notredame C. Multiple sequence alignment modeling: Methods and applications. *Briefings in Bioinformatics*, 2016. **17**(6): 1009–1023.

[38] Gotoh O. Multiple sequence alignment: Algorithms and applications. *Advances in Biophysics*, 1999. **36**: 159–206.

[39] Jiang Y., Liang Y., Wang D., Xu D., Joshi T. A dynamic programing approach to integrate gene expression data and network information for pathway model generation. *Bioinformatics*, 2020. **36**(1): 169–176.

[40] Gokhale M.B., Graham P.S. Reconfigurable Computing: Accelerating Computation with Field-programmable Gate Arrays, Springer Science & Business Media; 2006.

[41] Shikder R., Thulasiraman P., Irani P., Hu P. An OpenMP-based tool for finding longest common subsequence in bioinformatics. *BMC Research Notes*, 2019. **12**: 1–6.

[42] Lupyan D., Leo-Macias A., Ortiz A.R. A new progressive-iterative algorithm for multiple structure alignment. *Bioinformatics*, 2005. **21**(15): 3255–3263.

[43] Wallace I.M., Blackshields G., Higgins D.G. Multiple sequence alignments. *Current Opinion in Structural Biology*, 2005. **15**(3): 261–266.

[44] Can T. Introduction to bioinformatics. *miRNomics: MicroRNA Biology and Computational Analysis*, 2014. 51–71.

[45] Sharma V. Bioinformatics and its applications in environmental science and health and its applications in other disciplines. *Sambodhi Journal*, 2021. **4**(1): 88–93.

[46] Sharma V., Panwar A., Sharma A.K. Molecular dynamic simulation study on chromones and flavonoids for the in silico designing of a potential ligand inhibiting mTOR pathway in breast cancer. *Current Pharmacology Reports*, 2020. **6**: 373–379.

## Multiple choice questions

Q1 Which BLAST program is used to search for similar sequences in a nucleotide database?
    a) BLASTN
    b) BLASTP
    c) BLASTX
    d) TBLASTN

Q2 What is the purpose of the E-value in BLAST results?
    a) It represents the number of sequences in the database.
    b) It measures the significance of the sequence similarity.
    c) It indicates the length of the alignment.
    d) It represents the number of gaps in the alignment.

Q3 Which BLAST program is used to search for similar protein sequences in a protein database?
    a) BLASTN
    b) BLASTP
    c) BLASTX
    d) TBLASTN

Q4 What does the word "Local" refer to in the BLAST algorithm?
    a) It refers to the alignment of the entire sequence.
    b) It refers to searching for local similarities within sequences.
    c) It refers to searching for global similarities across sequences.
    d) It refers to the locality of the search server.

Q5   What is the primary purpose of sequence alignment in bioinformatics?
a) To identify homologous sequences
b) To predict protein structures
c) To perform gene editing
d) To synthesize DNA molecules

Q6   Which of the following algorithms is commonly used for pairwise sequence alignment?
a) Smith–Waterman
b) Needleman–Wunsch
c) BLAST
d) FASTA

Q7   Which type of sequence alignment is used to align more than two sequences simultaneously?
a) Global alignment
b) Local alignment
c) Multiple sequence alignment
d) Pairwise alignment

Q8   What is the purpose of gap penalties in sequence alignment algorithms?
a) To reward matches between sequences
b) To penalize mismatches between sequences
c) To reward the creation of gaps in alignments
d) To penalize the creation of gaps in alignments

Q9   Which of the following scoring matrices is commonly used in sequence alignment algorithms?
a) Smith–Waterman matrix
b) Needleman–Wunsch matrix
c) Blosum matrix
d) FASTA matrix

Q10  In pairwise sequence alignment, what does a positive score indicate?
a) A match between residues
b) A mismatch between residues
c) The presence of gaps in the alignment
d) The absence of gaps in the alignment

**Answers**
Q1   a)   BLASTN
Q2   b)   It measures the significance of the sequence similarity.
Q3   b)   BLASTP
Q4   b)   It refers to searching for local similarities within sequences.
Q5   a)   To identify homologous sequences
Q6   a)   Smith–Waterman
Q7   c)   Multiple sequence alignment
Q8   d)   To penalize the creation of gaps in alignments
Q9   c)   Blosum matrix
Q10  a)   A match between residues

Varruchi Sharma, Imran Sheikh, Vikas Kushwaha, Anil Panwar,
Seema Ramniwas, Anupam Sharma, Vandana Sharma, J. K. Sharma,
Sonal Datta, and Anil K. Sharma

# Chapter 4
# Tools used in sequence alignment

**Abstract:** In ancient times, when the number of organisms in this world was compara-
tively on the lower side, the evolution between them could be easily described. But as
newer organisms evolved, it became difficult to describe the evolutionary relationship
between them. To solve this problem, a branch of science called bioinformatics emerged.
Researchers developed a sequence alignment tool, which is a very important tool in today's era. Sequence alignment is a fundamental technique in bioinformatics that has
software programs to compare and align two or more biological sequences, such as
DNA, RNA, or protein sequences. These tools help researchers identify conserved regions,
functional domains, and evolutionary relationships among sequences. There are lots of
sequence alignment tools like ClustalW, Muscle, MAFFT, T-Coffee, BLAST (Basic Local
Alignment Search Tool), Needleman–Wunsch, and Smith–Waterman algorithm. They
are used by researchers to provide fruitful information for new research. These tools
typically accept input in various formats, such as FASTA or FASTQ, and provide output
in the form of alignment files (e.g., Clustal format) or visual representations. The choice
of alignment tool depends on the specific analysis requirements, the size of the dataset,
and the desired level of accuracy. Researchers often use a combination of tools and visual inspection to validate and interpret the results of sequence alignments. The actual
direction of sequence alignment tools will be influenced by ongoing research, technologi-
cal advancements, and the specific needs of the scientific and medical communities.

**Varruchi Sharma, Vikas Kushwaha**, Department of Biotechnology and Bioinformatics, Sri Guru
Gobind Singh College, Sector 26, Chandigarh 160019, India, e-mail: sharma.varruchi@gmail.com
**Imran Sheikh**, Department of Genetics, Plant Breeding and Biotechnology, Dr. K. S. Gill Akal College of
Agriculture, Eternal University, Baru Sahib, Sirmaur 173101, India
**Anil Panwar**, Department of Bioinformatics and Computational Biology, College of Biotechnology, CCS
Haryana Agricultural University, Hisar, Haryana, India
**Seema Ramniwas**, University Centre for Research and Development, University Institute of
Biotechnology, Chandigarh University, Gharuan, Mohali, India
**Anupam Sharma**, Department of Physics, Guru Kashi University, Talwandi Sabo, Bathinda, Punjab, India
**Vandana Sharma, J. K. Sharma**, Department of Physics, Maharishi Markandeshwar (Deemed to be
University), Mullana, Ambala 133207, Haryana, India
**Sonal Datta**, Department of Biosciences and Technology, Maharishi Markandeshwar (Deemed to be
University), Mullana, Ambala 133207, Haryana, India, e-mail: anibiotech18@gmail.com
**Anil K. Sharma**, Department of Biotechnology, Amity School of Biological Sciences, Amity University, Punjab
Mohali-140306, India

https://doi.org/10.1515/9783111568584-004

**Keywords:** Computational tools, sequence alignment, evolutionary relationship, BLAST, FASTA, ClustalW

There are some software tools that are used in pairwise and multiple sequence alignments (MSAs) at sequential and structural levels. Alignment tools perform all the functions associated with the similarity among DNA or protein sequences. For a long time, among the previous versions of the tools used for alignment, ClustalW had the wide acceptance for MSAs. In the present updated scenario, Clustal Omega is an improved version of the same tool.

There are various sequence alignment software, tools, and databases used in alignment (pairwise, MSA, genome-level analysis, etc.) (Table 4.1).

**Table 4.1:** Various sequence alignment tools and databases.

| S. no. | Name of the tool/database | Description | References |
|---|---|---|---|
| 1 | BLAST | Basic Local Alignment Search Tool is used for both protein and DNA sequence types | [1–3] |
| 2 | FASTA | DNA and protein sequence alignment software is also a text-based format used to represent nucleotide or amino acid (protein) sequences | [4] |
| 3 | PSI-BLAST | Position-specific iterative BLAST is quite more sensitive than BLAST | [5] |
| 4 | HMMER | Using hidden Markov models performs local and global searches using both DNA and protein sequence types | [6–9] |
| 5 | CUDASW++ | Compute Unified Device Architecture is an extension of C/C++ | [10, 11] |
| 6 | DIAMOND | Used for fast and sensitive protein alignment | [12] |
| 7 | SSEARCH | Comparison between both protein/DNA sequences and a protein/DNA sequence database using the Smith–Waterman algorithm | [13, 14] |
| 8 | ScalaBLAST | High-performance BLAST (rapid and robust calculations) | [15, 16] |
| 9 | SWIMM | Smith–Waterman Implementation for Intel Multicore and Manycore architectures | [17] |
| 10 | SAM | Sequence Alignment Map | [18, 19] |
| 11 | ALLALIGN | Aligns all sequences | |
| 12 | DNADot | Web-based dot-plot tool | [20] |
| 13 | LALIGN | Multiple, nonoverlapping, local similarity | [21–23] |

**Table 4.1** (continued)

| S. no. | Name of the tool/database | Description | References |
|---|---|---|---|
| 14 | mAlign | Modeling alignment | [24] |
| 15 | MUMmer | Comparing large genomes | [25–27] |
| 16 | PyMOL | Sequence and structure-based sequence alignment | [28, 29] |
| 17 | Needle | Needleman–Wunsch dynamic programming | [30] |
| 18 | UGENE | For the management, analysis, and visualization of data | [31] |
| 19 | SEQALN | Dynamic programming in both local and global alignments | [32] |
| 20 | AMAP | Sequence annealing | [33] |
| 21 | ClustalW | Progressive alignment | [34, 35] |
| 22 | Compass | Comparison of multiple protein sequence alignments with an assessment of statistical significance | [36] |
| 23 | DNADynamo | DNA sequence analysis software package | [37] |
| 24 | Geneious | A progressive-iterative alignment tool | [38] |
| 25 | DNASTAR | Lasergene Molecular Biology Suite | [39] |
| 26 | MUSCLE | Program of multiple alignments | [37] |
| 27 | T-Coffee | MSA | [40] |
| 28 | SAGA | Sequence alignment by genetic algorithm | [41] |
| 29 | MGA | Multiple Genome Aligner | [42] |
| 30 | BLAT | BLAST-like alignment tool | [3, 43] |
| 31 | BFAST | Alignment tool for large-scale genomic resequencing | [44] |
| 32 | AVID | Whole-genome alignment using the pairwise global method | [45] |

Data retrieval based on similarity is the main implication of pairwise alignment. In the case of biological data retrieval, the process is followed by the involvement of certain components such as the requirement of a query; i.e., a query sequence is needed, and the comparison among databases is performed. After performing a similarity comparison, the statistical assessment of data is carried out, which addresses the significance or accuracy of the query sequence to that of the database sequence and also infers the information at the homology level transmitted to the query sequence. Further, we have tried to explain some important methods used in bioinformatics for performing sequence alignment.

Pairwise local sequence alignment tools are used to identify the regions of sequential similarity/identity among pairs of either DNA/RNA or among protein sequences.

EMBOSS Water: The alignments are performed using the Smith–Waterman algorithm for both nucleotide and protein sequences using the following formats: GCG, FASTA, EMBL, GenBank, PIR, NBRF, PHYLIP, or UniProtKB/Swiss-Prot format [46].

EMBOSS Matcher: Pairwise sequence alignment of both DNA and protein sequences is performed based on the LALIGN application using the following formats: GCG, FASTA, EMBL, GenBank, PIR, NBRF, PHYLIP, or UniProtKB/Swiss-Prot format [46].

## 4.1 Multiple sequence alignment

The alignment of similar sequence regions among three or more DNA/RNA or protein sequences can be performed using many programs such as ClustalW [34], a widely accepted tool that has been used successfully for many years. By this time, Clustal Omega, another improved version of the tool, is available for the work to be carried out in a successful and effective manner. The same tool can be used in the formats GCG, FASTA, EMBL, GenBank, PIR, NBRF, PHYLIP, or UniProtKB/Swiss-Prot, along with the acceptance of up to a maximum of 2,000 sequences. This can be used online via Clustal Omega (EMBL-EBI with FAQ (Help) Page) web server [47–49]. T-Coffee, another program of MSA, which stands for tree-based consistency objective function used for alignment evaluation, is an iterative MSA algorithm [40] (Table 4.2).

In bioinformatics, there are some tools and software programs for data extraction. The most understanding, well-known and widely used program is BLAST along with a full BLAST suite of programs. We can run these programs locally, as there are some web pages that allow users to compare a protein or DNA sequence.

**Table 4.2:** Various Tools and Webpages for sequence comparison.

| S. no. | Tools | Web pages | References |
|---|---|---|---|
| 1 | Web pages for the comparison of a protein or DNA sequence | National Center for Biotechnology Information (USA) Searches European Bioinformatics Institute (UK) Searches | [50] |
| | | BLAST search through SBASE (domain database; ICGEB, Trieste) | |
| 2 | Approaches for matching a single sequence to a database included | The FASTA suite (William Pearson, University of Virginia, USA) | |
| | | SCANPS (Geoff Barton, European Bioinformatics Institute, UK) | |
| | | BLITZ (Compugen's fast Smith–Waterman search) | |

**Table 4.2** (continued)

| S. no. | Tools | Web pages | References |
|--------|-------|-----------|------------|
| 3 | Multiple sequence information | PSI-BLAST (NCBI, Washington) | |
| | | ProfileScan Server (ISREC, Geneva) | |
| | | HMMER hidden Markov model searching (Sean Eddy, Washington University) | |
| | | Wise package (Ewan Birney, Sanger Center; this is for protein versus DNA comparisons) | |

# 4.2 BLAST

BLAST is a tool used for finding the best similarity searches among sequences (in DNA or protein sequences). It is a program that is most popular around the world as it is available through the most popular server, the National Center for Biotechnology Information (NCBI) [50]. The program estimates alignments that enhance or find the extent of local similarity and also helps in finding the maximal segment pair (MSP) score. The MSP analyzes the statistical significance and enactment of the alignments generated by the program. BLAST works upon the heuristic approach/method, which involves finding interrelated sequences in the database search or shortcuts to get the accurate response of alignment in a much faster way. The program targets short conjoint patterns in the query and database sequences and further joins these queries, and the search is extracted into an alignment as FASTA works. Both BLAST and FASTA are alike, but BLAST emphasizes substantial patterns in DNA or protein sequences [51, 52].

## 4.2.1 Working of BLAST

BLAST works on the strategy of finding ungapped segments that have the best score among related sequences. The following steps are performed during the BLAST search:

It may first check the query for low complexity or other repeats and then produce a set of "words" – short, fixed-length sequences based on the query. They are used to initiate matches in the database (or "subject") sequences. In the preliminary search, a number of steps are performed on every sequence in the database. First, the database is scanned for matches to the words, and those are used to initiate a gap-free extension. Second, gap-free extensions that achieve a certain score are used to seed a gapped extension that only calculates the score and extent and leaves to a later stage the time- and memory-consuming work of calculating insertions and deletions. Gapped extensions that achieve a specified score are saved, though lower scoring matches may be deleted if too many matches are found. In the final traceback phase of the search, gapped extensions saved in the preliminary phase are used as seeds for a gapped extension that also calculates the insertions and deletions and may use more sensitive parameters.

## 4.2.2 Programs of BLAST

There are five programs with which we can perform BLAST searches for our desired results: BLASTn, BLASTp, BLASTx, tBLASTn, and tBLASTx. BLASTn and BLASTp are used for comparing nucleotide or protein sequences [53] (Figure 4.1 and Table 4.3).

**Figure 4.1:** BLAST Programs.

**Table 4.3:** Various BLAST programs.

| S. no. | BLAST program | Query sequence | Database used | Objective | References |
|---|---|---|---|---|---|
| URL: https://blast.ncbi.nlm.nih.gov/Blast.cgi | | | | | |
| 1 | BLASTp | Protein sequence | Against protein sequence database | Characterizes mutual regions among proteins | [54, 55] |

**Table 4.3** (continued)

| S. no. | BLAST program | Query sequence | Database used | Objective | References |
|---|---|---|---|---|---|
| 2 | BLASTn | Nucleotide sequence | Against nucleotide database | Oligonucleotide mapping Genomic sequence annotation Sequencing pattern assembly | |
| 3 | BLASTx | Nucleotide translated into protein sequence | Against protein database | Protein-coding regions of genes in genomic DNA | |
| 4 | tBLASTn | Protein sequence | Nucleotide translated into protein | Identifying transcripts, potentially from multiple organisms, similar to a given protein; mapping a protein to genomic DNA | |
| 5 | tBLASTx | Nucleotide translated into protein sequence | Nucleotide translated into protein | Cross-species gene prediction at the genome or transcript level; searching for genes missed by traditional methods or not yet in protein databases | |
| 6 | PSI-BLAST | Accepts any number of protein sequences as input. | Protein databases | It performs a preliminary BLASTP search in order to collect significant information and then uses to produce a position-specific scoring matrix (PSSM) | |
| 7 | DELTA-BLAST | Protein sequence database using a PSSM constructed from conserved domains matching a query | | Produces a PSSM with a fast RPSBLAST search of the query, followed by searching this PSSM against a database of protein sequences | |

## 4.2.3 BLAST results

The BLAST result provides the information about the query sequence we used along with the database information, taxonomy information, graphical summary of results, and the descriptions of alignments. The alignment's primary purpose is to deduce homology. There is a separate complimentary section for conserved domains, which provides a graphical overview of the conserved domains. Another section consists of a graphical overview that shows the alignment of subject sequences to that of the query provided. All the hits are represented in this window by colored bars below the query sequence. Clicking on a particular hit takes the user to the pairwise alignment among hit and query sequence, and the bar color of a particular hit refers to the alignment score [54]. The text summary of the sequence is provided below the graphical over-

| | Description | Max Score | Total Score | Query Cover | E value | Per. Ident | Accession |
|---|---|---|---|---|---|---|---|
| ☑ | serine/threonine-protein kinase mTOR [Homo sapiens] | 5307 | 5307 | 100% | 0.0 | 100.00% | NP_004949.1 |
| ☑ | serine/threonine-protein kinase mTOR [Pan paniscus] | 5306 | 5306 | 100% | 0.0 | 99.96% | XP_003822131.1 |
| ☑ | serine/threonine-protein kinase mTOR [Gorilla gorilla gorilla] | 5305 | 5305 | 100% | 0.0 | 99.96% | XP_030867095.1 |
| ☑ | serine/threonine-protein kinase mTOR [Pongo abelii] | 5304 | 5304 | 100% | 0.0 | 99.96% | XP_024108864.1 |
| ☑ | FRAP1 variant protein [Homo sapiens] | 5303 | 5303 | 100% | 0.0 | 99.96% | BAE06077.1 |
| ☑ | serine/threonine-protein kinase mTOR isoform X1 [Nomascus leucogenys] | 5303 | 5303 | 100% | 0.0 | 99.92% | XP_030661600.1 |
| ☑ | PREDICTED: serine/threonine-protein kinase mTOR [Macaca fascicularis] | 5303 | 5303 | 100% | 0.0 | 99.88% | XP_005544842.1 |

**SIGNIFICANT ALIGNMENT**

*Score:* Using the scoring matrix and gap penalties, the values computed are defined as score. A high score means maximum similarity

*Query Coverage:* It's the percentage of our sequence in the alignment that is aligned to a sequence in GenBank

*E-Value: Expect Value is defined as the* number of hits "expected" by chance in searching a database of a certain size

*Percent identity:* Similarity measurement among two DNA/protein sequences

*Hits:* Alignments among query and subject sequence

*Bits: The standardized* score is monitored by bits

*Accession:* Matchless identification number

**DESCRIPTIONS**

Alignment Scores  ■ < 40  ■ 40 - 50  ■ 50 - 80  ■ 80 - 200  ■ >= 200

**Distribution of the top 100 Blast Hits on 100 subject sequences**

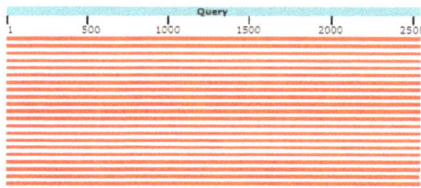

Query
1   500   1000   1500   2000   2500

← **GRAPHICAL REPRESENTATION**

**GRAPHIC SUMMARY**

Range 1: 1 to 2549 GenPept Graphics

| Score | Expect | Method | Identities | Positives | Gaps |
|---|---|---|---|---|---|
| 5307 bits(13767) | 0.0 | Compositional matrix adjust. | 2549/2549(100%) | 2549/2549(100%) | 0/2549(0%) |

```
Query  1    MLGTGPAAATTAATTSSNVSVLQQFASGLKSRNEETRAKAAKELQHVVTMELREHSQEES  60
            MLGTGPAAATTAATTSSNVSVLQQFASGLKSRNEETRAKAAKELQHVVTMELREHSQEES
Sbjct  1    MLGTGPAAATTAATTSSNVSVLQQFASGLKSRNEETRAKAAKELQHVVTMELREHSQEES  60

Query  61   TRFYDQLNHHIFELVSSSDANERKGGILAIASLIGVEGGNATRIGRFANYLRNLLPSNDP  120
            TRFYDQLNHHIFELVSSSDANERKGGILAIASLIGVEGGNATRIGRFANYLRNLLPSNDP
Sbjct  61   TRFYDQLNHHIFELVSSSDANERKGGILAIASLIGVEGGNATRIGRFANYLRNLLPSNDP  120
```

**ALIGNMENT RESULT**

*Identities:*

*Positives:*

*Gaps:*

*Query Sequence: Sequence submitted for* alignment

*Subject Sequence:* Sequence that we get from the database after alignment

**ALIGNMENT**

**Figure 4.2:** Alignment and its procedure.

view bar, which shows the best alignment hits of the alignments with the query sequence, and also the significant alignments are arranged from top to down. The alignments represented are based on the following criteria: score or bit score (value calculated from the number of gaps and substitutions that is having association with each aligned sequence), high score represents best/significant alignment, E-value (expect value – the probability that a sequence with a similar score will occur in the database by chance, and smaller the E-value, more significant/best is the alignment), query coverage, and percent identity. The taxonomic view of results shows a report about the organism, lineage, and the taxonomy of the organism [1, 54, 56] (Figure 4.2).

Some BLAST-related tools are as follows:

Primer-BLAST: It is used for finding out the primers that amplify only a specific gene [57].

IgBLAST: This type annotates the flexible regions of an immunoglobulin sequence that also includes a variable (V), diversity (D), and a joining (J) segment [58].

VecScreen: This helps in identifying the vector contamination in a query sequence.

MSA is an essential and most accepted computational method that is used for sequence analysis. In MSA, the homologous sequences are compared for performing various biological functions such as for the construction of phylogenetic trees and for the analysis of protein secondary and tertiary structure. This can later help in identifying new members of protein families, to find out closely related genes/proteins for identifying the evolutionary relationships among genes.

There are a number of MSA algorithms with an improved computational complexity that are also more accurate for the alignments.

ClustalW is one of the most popular algorithms that belongs to the Clustal family, which has more incorporation to commercially available bioinformatic packages used for alignment purposes. Nowadays, Clustal Omega has gained more attraction in the same process of alignment, and we also have T-Coffee, MUSCLE, Kalign, and Mafft MSA algorithms.

ClustalW (weights) was introduced by Thompson et al. It was considered to be the best method in terms of alignment accuracy/quality, sensitivity, and speed in contrast to other algorithms [59]. The method performs pairwise alignment using the k-tuple method. In the same process, the calculation of a matrix is performed, which demonstrates the similarity of each pair of sequences. The similarity scores obtained are transformed into distance scores, and then by using the algorithm the distance scores produce a guide tree (neighbor-joining (NJ) method) (Figure 4.3).

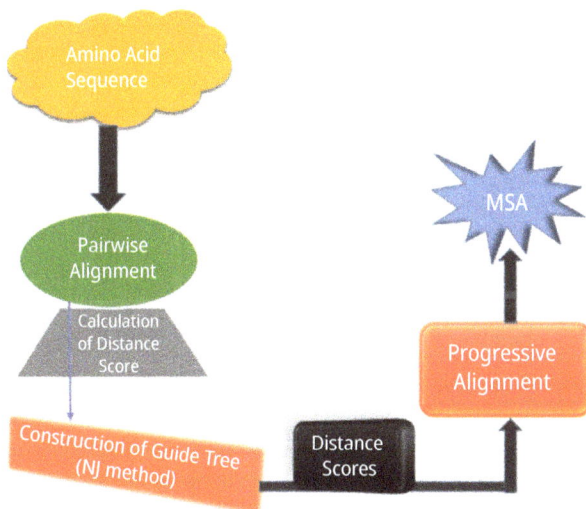

**Figure 4.3:** Demonstration of ClustalW algorithm.

The NJ method keeps a check on the nodes on a tree than that of a taxa or clusters of taxa. This program is a general-purpose MSA program for either DNA or proteins. The sequences used must be in a file in any one of the formats out of the following: FASTA " >," PIR, Swiss-Prot, Clustal (*.aln), GCG-MSF (Pileup), GCG9-RSF, and GDE flat file. In continuation to the process, the program identifies different file formats used and also guesses the sequence type, whether it is an amino acid or nucleotide sequence. After the sequences have been loaded for the purpose of MSA, we will be encouraged for the alignment result and to store a dendrogram (which shows the similarity of the sequences to one another). The alignment performed here is carried out in three steps: (a) pairwise alignment is performed (speed sensitivity), (b) dendrogram is constructed, and (c) in the last step the multiple alignment is carried out. The output of the alignment performed is provided in any one of the following formats: FASTA, CLUSTAL, NBRF-PIR, GCG, GDE, PHYLIP, NEXUS, etc. For the score calculation of the alignments performed, the following two methods are used:

(a) Dynamic programming: This is a slow-accurate method and it is best considered for short sequences.
(b) Wilbur and Lipman algorithm: This is also called as NUCALN, generated by Wilbur and Lipman for the quick identification of sequence similarities. The algorithm is based on the identification of tuples of k nucleotides (or amino acid residues) that are the same in the two sequences.

After the calculation of alignment, the phylogenetic trees are constructed. Two main methods are used for tree construction: (i) NJ and (ii) unweighted pair group method with arithmetic mean (UPGMA). For tree construction, first the distances among alignments are calculated between all pairs of sequence from multiple alignments; second, NJ or UPGMA method is applied for the distance matrix (Figure 4.4).

## 4.3 T-Coffee

Tree-based consistency objective function for alignment evaluation (T-Coffee) is another progressive algorithm, which has a focus on solving the errors introduced by the progressive methods. This method also takes into account the pairwise alignments (Figure 4.5).

This method in the first step creates a library (using an algorithm that is used in ClustalW). In this constructed library, additional information is added, such as composition of protein domains and structural information, which further leads to the creation of progressive alignment. T-Coffee method has a limitation as this is quite slower than that of ClustalW, and it cannot work with more than a few hundred sequences (Figure 4.6).

```
CLUSTAL 2.1 multiple sequence alignment
ABC46896.1       MTEYKLVVVGAGGVGKSALTIQLIQNHFVEEFDPTIEDSYRKQVVIDGETCLLDILDTAG
ABA82136.1       MTEYKLVVVGAGGVGKSALTIQLIQNHFVEEYDPTIEDSYRKQVVIDGETCLLDILDTAG
ABA82135.1       MTEYKLVVVGAGGVGKSALTIQLIQNHFVDEYDPTIEDSYRKQVVIDGETCLLDILDTAG
                 *****************************.*.****************************

ABC46896.1       QEEYSAMRDQYMRTGEGFLCVFAVNNTKSFEDINQYREQIKRVKDADEVPMVLVGNKVDL
ABA82136.1       QEEYSAMRDQYMRTGEGFLCVFAVNNLKSFEDINQYREQIKRVKDADEVPMVLVGNKVDL
ABA82135.1       QEEYSAMRDQYMRTGEGFLCVFAINNTKSFEDIHHYREQIKRVKDSEDVPMVLVGNKCDL
                 ***********************.** ******.:***********.::*********.**

ABC46896.1       PTRTVDAKQARPVADSYNIPYVETSAKTRQGVDDAFYTLVREIRKYKERKGP--------
ABA82136.1       PIRNVEQRQGKHMAELYHIPYVETSAKTRQGVDDAFYTLVREIRKYKEKKGKKEKKRKTG
ABA82135.1       PSRTVDTKQAQDLARNYGIPFIETSAKTRQGVDDAFYTLVREIRKHKEKTSKEGRKKKKK
                 * *.*: :*.: :*   *  **::****************************.**:..

ABC46896.1       ---------
ABA82136.1       KRSCELL--
ABA82135.1       KSKAKCVVM
```

Phylogram   (midpoint rooted tree)
▢ without branch length   ▢ without branch length labels   ▢ without leaf labels   ▢ without ticks
JSON⬇  SVG⬇  PNG⬇
         0     0.2   0.4   0.6   0.8    1    1.2   1.4

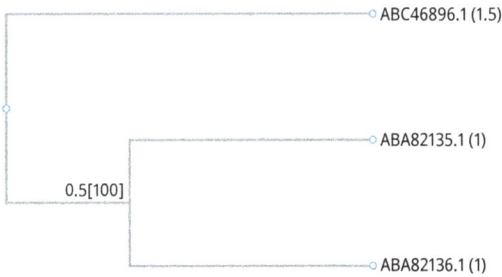

**Figure 4.4:** Result showing ClustalW alignment in the top and tree (phylogram) constructed showing branch length and labels.

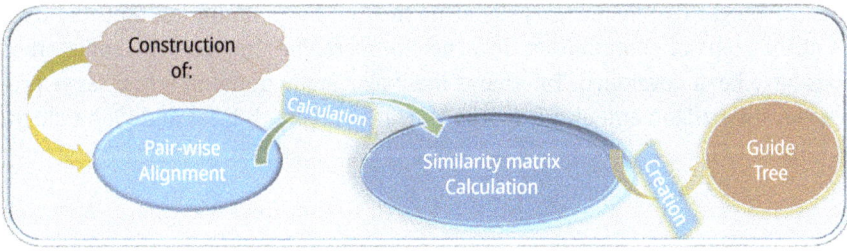

**Figure 4.5:** Steps followed by the T-Coffee algorithm.

## T-Coffee alignment result

### MSA

*The multiple sequence alignment result as produced by T-coffee*

```
T-COFFEE, Version_11.00.d625267 (2016-01-11 15:25:41 - Revision d625267 - Build 507)
Cedric Notredame
SCORE=999
*
BAD AVG GOOD
*
ABC46896.1   : 100
ABA82136.1   :  99
AHA93923.1   :  99
cons         :  99

ABC46896.1   MTEYKLVVVGAGGVGKSALTIOLIONHFVEEFDPTIEDSYRKOVVIDGETCLLDILDTAGOEEYSAMRD
ABA82136.1   MTEYKLVVVGAGGVGKSALTIOLIONHFVEEYDPTIEDSYRKOVVIDGETCLLDILDTAGOEEYSAMRD
AHA93923.1   MTEYKLVVVGAGGVGKSALTIQLIQNHFVDEYDPTIEDSYRKQVVIDGETCLLDILDTAGQEEYSAMRD

cons         *****************************.*:*********************************

ABC46896.1   OYMRTGEGFLCVFAVNNTKSFEDINOYREOIKRVKDADEVPMVLVGNKVDLPTRTVDAKOARPVADSYN
ABA82136.1   OYMRTGEGFLCVFAVNNLKSFEDINOYREOIKRVKDADEVPMVLVGNKVDLPIRNVEOROGKHMAELYH
AHA93923.1   QYMRTGEGFLLVFAVNNAKSFEDISAYREQIKRVKDADVVPMVLVGNKCDLQVRAMDMQQAREVAKNYD

cons         ********** ****** ******. *********** ********* ** * :: :*.: :*. *.

ABC46896.1   IPYVETSAKTROGVDDAFYTLVREIRKYKERKGP----------------
ABA82136.1   IPYVETSAKTROGVDDAFYTLVREIRKYKEKKGKEKKRKTGKRSCELL
AHA93923.1   IPFIETSAKTRMGVDDAFYTLVREIRKDRERRGKVRIGSRNSKRKCIVF

cons         **:;******* ***************** :*::*
```

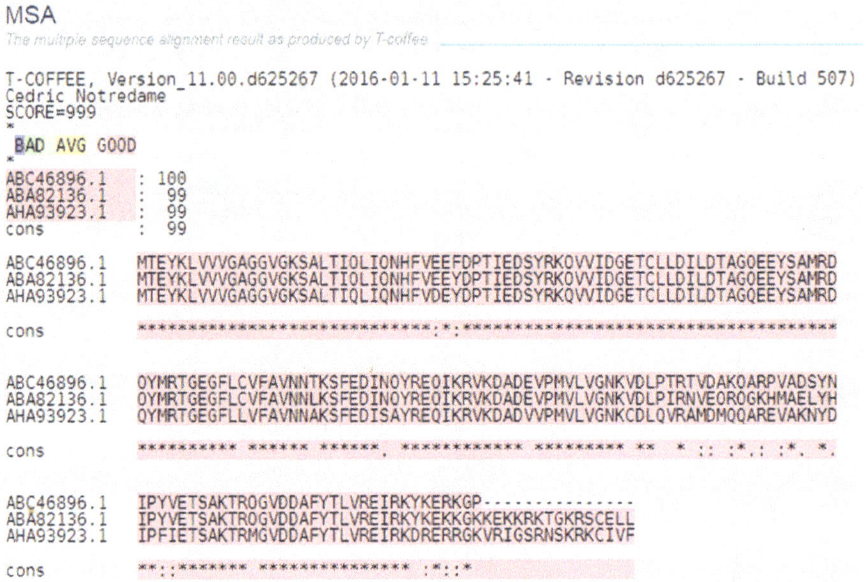

**Figure 4.6:** T-Coffee-alignment result.

MSA is the most significant and most essential technique in bioinformatics and its applications. For performing alignments, a number of algorithms have been developed, where some of them are either exhaustive (less feasible) or heuristic (most commonly used) in nature. There are three main types of approaches used in alignment, that is, progressive, iterative, and block based. The progressive method works in pairwise similarity methods; an example of such a method is Clustal, which works with an application of weighting scheme, and the same method has a limitation that this fixes errors at the steps of computation. In order to overcome the problem, some other methods have been developed. T-Coffee is the major method that in a real sense provides a sensitive alignment, and it works by combining both local and global alignments (Figure 4.6).

## 4.4 HMMER

HMMER is a well-known bioinformatics tool that offers a web server and a command line tool to users for searching sequence databases for sequence homologs, and for making sequence alignments using probabilistic methods called profile hidden Markov models (profile HMMs) (http://hmmer.org; accessed on June 20, 2023). This tool was developed by Eddy et al. in 1995. The HMMER software suite has been widely used, particularly by protein family databases such as Pfam and InterPro and their associated search tools. In 2010, a newer version HMMER 3.0 with a 100-fold speed gains relative to previous versions or BLASTP [60]. This version has been upgraded to the latest version of the HMMER software, version 3.1b2, that includes minor bug fixes but, more importantly, has additional performance improvements over the previous version, 3.0 [61]. The initial release of the website provided the following three search algorithms:

- *phmmer*: Single protein sequence against protein sequence database
- *hmmscan*: Single protein sequence against profile HMM library (Pfam)
- *hmmsearch*: Either MSA or profile HMM against a protein sequence database

## 4.4.1 Working with HMMER (https://www.ebi.ac.uk/Tools/ hmmer/) (Figure 4.7)

- Homology search can be easily performed using HMMER. If you are working with protein sequences, then use PHMMER.
- Build a local database of your own and then perform a homology search within it using a query sequence.
- Add column masks to MSAs using the *alimask* module.
- Align sequences to a profile.
- Construct profiles from MSAs.
- Convert profile files to various formats.
- Retrieve profiles from a file.
- Generate a conversation logo from a profile.
- Search sequences against a profile and sequence database.
- Search DNA/protein queries against profile and sequence database.
- You can also perform a homology search iteratively.
- Translate DNA into six frames.
- Calculate sequence weights in MSAs.
- Mask sequence residues.
- In addition to the creation of consensus secondary structures, you can also calculate the accuracy of predicted RNA secondary structures using HMMER.
- Reverse compliment an MSA.
- Map two alignments to each other.

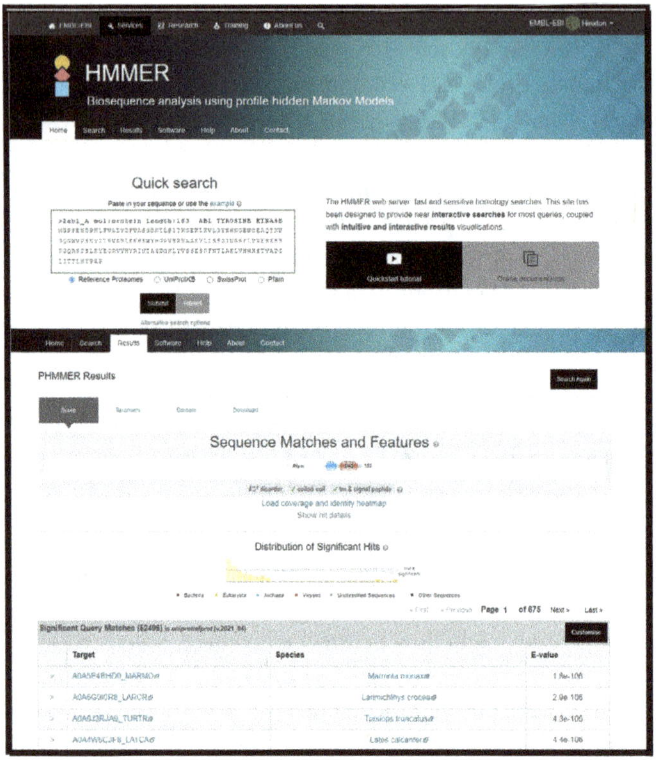

**Figure 4.7:** HMMER user interface.

## 4.5 DIAMOND

DIAMOND (double index alignment of next-generation sequencing data) is a sequence aligner for protein and translated DNA searches designed for high-performance analysis of big sequence data. DIAMOND is a fast and sensitive protein aligner that was initially developed for metagenomic applications to achieve ultrafast alignments at the cost of alignment sensitivity, compared with the gold standard, BLAST. DIAMOND is available as open-source software under the GPL3 license (http://www.diamond search.org) [62].

DIAMOND uses double indexing, an approach that determines the list of all seeds and their locations in both query and reference sequences. The two lists are sorted lexicographically and traversed together in a linear manner to determine all matching seeds and their corresponding locations. Double indexing takes advantage of the cache hierarchy by increasing data locality, thus reducing the demands on main memory bandwidth [63].

The key features are as follows:

- Pairwise alignment of proteins and translated DNA at 100× to 10,000× speed of BLAST
- Protein clustering of up to tens of billions of proteins
- Frameshift alignments for long read analysis
- Low-resource requirements and suitable for running on standard desktops or laptops
- Various output formats, including BLAST pairwise, tabular, and XML, as well as taxonomic classification

## 4.6 ScalaBLAST

ScalaBLAST uses a high-performance multiprocessor system that utilizes the NCBI BLAST library and supports all five primary program types (blastn, blastp, tblastn, tblastx, and blastx) with several output formats (pairwise, tabular, or XML). Scala-BLAST is designed to run a large number of queries against large or small databases. The main limitation of ScalaBLAST version 1.0 was the use of static data partitioning that did not have fault-resilience properties. Further, these limitations in ScalaBLAST 2.0 were resolved by (i) re-implementing the task scheduling layer by introduction of a dynamic task management scheme that (ii) does not require preformatting [64]. Sca-laBLAST 2.0 can be downloaded freely from http://omics.pnl.gov/software/ScalaBLAST. php. On both small- and large-scale systems, scalaBLAST 2.0 allows users to accelerate the throughput of BLAST calculations that are complete even when processes fail in support of robust sequence analysis applications.

## 4.7 Sequence alignment/map

Sequence alignment/map (SAM) format is a generic alignment format for storing read alignments against reference sequences, supporting short and long reads (up to 128 Mbp) produced by different sequencing platforms. It supports single- and paired-end reads and combines reads of different types, including color space reads from the AB/SOLiD sequencing tool. It is designed to scale to alignment sets of $10^{11}$ or more base pairs, which is typical for the deep resequencing of one human individual. The SAM format consists of one header section and one alignment section. The lines in the header section start with the character "@," and the lines in the alignment section do not [65].

## 4.8 MALIGN

MALIGN is a library for performing multiple alignments on sequences of different domains, allowing the usage of asymmetric scoring matrices. Multiple alignments are actual multiple alignments, scoring according to the overall probability of each alignment site, and not a succession of pairwise alignments gradually combined. MALIGN employs an optimality criterion of a minimal cost of phylogenetic tree to choose the best alignment. This phylogenetic cost involves the calculation of minimum length spanning trees and a variety of options, from the most simple heuristics, to those with several forms of branch swapping, to exact branch-and-bound solutions, are provided.

Either a command line or an interactive interface can be used to run the program. The specification of a parameter file and entering the parameters directly from the command line are both acceptable ways to enter the analysis parameters. A modified GenBank sequence format serves as the basis for the input data file format. Nucleic acid and amino acid sequences can freely mix, and MALIGN will transform them into nucleotide ambiguity representations. The numerals 0–9 may also be given triplet equivalents, and amino acid labels may be changed (for various triplet codes) [66].

## 4.9 COMPASS

COMPASS (comparison of multiple protein alignments with assessment of statistical significance) method involves the construction of local profile–profile alignments allowing gaps by means of a dynamic programming algorithm. The two approaches FFAS17 and prof_sim for building local profile–profile alignments have both been documented. These two techniques contain the scoring frameworks for alignment creation and the protocols for generating profiles from numerous alignments. The "dot-product" scores, which are connected to the correlation coefficients between the amino acid frequencies within the two columns, are used by FFAS to evaluate how similar the profile columns are to one another. A more complex scoring method is used by the program, which computes the divergence score and the significance score by applying the Jensen–Shannon measure to the divergence between two probability distributions [67].

## 4.10 Conclusion

In conclusion, sequence alignment tools play a crucial role in the field of bioinformatics and genomics, enabling researchers to compare biological sequences, identify functional elements, and gain insights into the underlying genetic code. They have revolutionized our understanding of biology and continue to be a driving force in unlocking the secrets of life. As technology and research progress, we can anticipate even more sophisticated and efficient tools that will further enhance our ability to analyze and interpret the wealth of genomic data at our disposal.

# References

[1]     Korf I., Yandell M., Bedell J. Blast, " O'Reilly Media, Inc."; 2003.
[2]     Altschul S.F., et al. Basic local alignment search tool. *Journal of Molecular Biology*, 1990. **215**(3): 403–410.
[3]     Kent W.J. BLAT – The BLAST-like alignment tool. *Genome Research*, 2002. **12**(4): 656–664.
[4]     Pearson W.R. Finding protein and nucleotide similarities with FASTA. *Current Protocols in Bioinformatics*, 2016. **53**: 3.9.1–3.9.25.
[5]     Altschul S.F., et al. Gapped BLAST and PSI-BLAST: A new generation of protein database search programs. *Nucleic Acids Research*, 1997. **25**(17): 3389–3402.
[6]     Durbin R., et al. Biological Sequence Analysis: Probabilistic Models of Proteins and Nucleic Acids, Cambridge University Press; 1998.
[7]     Finn R.D., et al. HMMER web server: 2015 update. *Nucleic Acids Research*, 2015. **43**(W1): W30–W38.
[8]     Eddy S. HMMER user's guide. Department of Genetics. *Washington University School of Medicine*, 1992. **2**(1): 13.
[9]     Potter S.C., et al. HMMER web server: 2018 update. *Nucleic Acids Research*, 2018. **46**(W1): W200–W204.
[10]    Nickolls J., et al. Scalable parallel programming with CUDA. *ACM Queue*, 2008. **6**(2): 40–53.
[11]    Liu Y., Maskell D.L., Schmidt B. CUDASW++: Optimizing Smith-Waterman sequence database searches for CUDA-enabled graphics processing units. *BMC Research Notes*, 2009. **2**: 73–73.
[12]    Buchfink B., Xie C., Huson D.H. Fast and sensitive protein alignment using DIAMOND. *Nature Methods*, 2015. **12**(1): 59–60.
[13]    Manavski S.A., Valle G. CUDA compatible GPU cards as efficient hardware accelerators for Smith-Waterman sequence alignment. *BMC Bioinformatics*, 2008. **9**(S2): S10.
[14]    Rognes T., Seeberg E. Six-fold speed-up of Smith–Waterman sequence database searches using parallel processing on common microprocessors. *Bioinformatics*, 2000. **16**(8): 699–706.
[15]    Oehmen C.S., Baxter D.J. ScalaBLAST 2.0: Rapid and robust BLAST calculations on multiprocessor systems. *Bioinformatics*, 2013. **29**(6): 797–798.
[16]    Oehmen C., Nieplocha J. ScalaBLAST: A scalable implementation of BLAST for high-performance data-intensive bioinformatics analysis. *IEEE Transactions on Parallel and Distributed Systems*, 2006. **17**(8): 740–749.
[17]    Rucci E., et al. SWIMM 2.0: Enhanced Smith–Waterman on Intel's multicore and manycore architectures based on AVX-512 vector extensions. *International Journal of Parallel Programming*, 2019. **47**(2): 296–316.
[18]    Li H., et al. The sequence alignment/map format and SAM tools. *Bioinformatics*, 2009. **25**(16): 2078–2079.
[19]    Hughey R., Karplus K., Krogh A. SAM: Sequence Alignment and Modeling Software System., *Technical Report UCSC-CRL-99-11*, 2003.
[20]    Celamkoti S., et al. GeneOrder3.0: Software for comparing the order of genes in pairs of small bacterial genomes. *BMC Bioinformatics*, 2004. **5**: 52.
[21]    Pearson W. LALIGN-Find multiple matching subsegments in two sequences. 2015. **28**(10).
[22]    Huang X., Miller W. LAlign (v 36.3. 6) finds non-overlapping local alignments. *Advances in Applied Mathematics*, 1991. **12**: 373–381.

[23] Lakhani J., Khunteta A., Choudhary A., Harwani D. MPSAGA: A matrix-based pair-wise sequence alignment algorithm for global alignment with position based sequence representation. *Sādhanā*, 2019. **44**: 1–17.

[24] Wheeler W., Gladstein D. MALIGN: A multiple sequence alignment program. *Journal of Heredity*, 1994. **85**(5): 417–418.

[25] Marçais G., et al. MUMmer4: A fast and versatile genome alignment system. *PLoS Computational Biology*, 2018. **14**(1): e1005944.

[26] Moore G. Mummer: Worship. *Universal Review*, 1888. **2**(5): 105–118.

[27] Kurtz S., et al. Open source MUMmer 30 is described in versatile and open software for comparing large genomes. *Genome Biology*, 2004. **5**: R12.

[28] Westin C., Hanson B., Craig P. Using PyMOL's align feature to create a database of ligand binding site files for the structural analysis of proteins. *FASEB Journal*, 2007. **21**: A296. doi: 10.1096/fasebj.21.5.A296-d.

[29] DeLano W.L. PyMOL Reference Guide, San Carlos, CA, USA: DeLano Scientific; 2004.

[30] Likic V. The Needleman-Wunsch Algorithm for Sequence Alignment. In *Lecture Given at the 7th Melbourne Bioinformatics Course, Bi021 Molecular Science and Biotechnology Institute*, University of Melbourne; 2008, 1–46.

[31] Okonechnikov K., et al. Unipro UGENE: A unified bioinformatics toolkit. *Bioinformatics*, 2012. **28**(8): 1166–1167.

[32] Gardner-Stephen P., Knowles G. Dash: Localising dynamic programming for order of magnitude faster, accurate sequence alignment. In *Proceedings. 2004 IEEE Computational Systems Bioinformatics Conference, 2004 (CSB 2004)*, 2004. IEEE.

[33] Csűrös M., Juhos S., Bérces A. Fast Mapping and Precise Alignment of AB SOLiD Color Reads to Reference DNA. In International Workshop on Algorithms in Bioinformatics, Springer; 2010.

[34] Thompson J.D., Gibson T.J., Higgins D.G. Multiple sequence alignment using ClustalW and ClustalX. *Current Protocols in Bioinformatics*, **2003**(1): 2.3.1–2.3.22.

[35] Li K.-B. ClustalW-MPI: ClustalW analysis using distributed and parallel computing. *Bioinformatics*, 2003. **19**(12): 1585–1586.

[36] Sadreyev R., Grishin N. COMPASS: A tool for comparison of multiple protein alignments with assessment of statistical significance. *Journal of Molecular Biology*, 2003. **326**(1): 317–336.

[37] Edgar R.C. MUSCLE: A multiple sequence alignment method with reduced time and space complexity. *BMC Bioinformatics*, 2004. **5**(1): 113.

[38] Iliyas S., Sarkhwas F. A comparative study of MSA tools based on sequence Alignment feachers and platform Independency to select the appropriate tool desired. *International Journal of Data Mining and Bioinformatics*, 2011. 66–72.

[39] Xia Z., Tan X., Zhang L. Prediction and sequence alignment of fruit dehiscence-related genes in oilseed rape. In *2008 2nd International Conference on Bioinformatics and Biomedical Engineering*, 2008 May 16, 97–100.

[40] Notredame C., Higgins D.G., Heringa J. T-Coffee: A novel method for fast and accurate multiple sequence alignment. *Journal of Molecular Biology*, 2000. **302**(1): 205–217.

[41] Notredame C., Higgins D.G. SAGA: Sequence alignment by genetic algorithm. *Nucleic Acids Research*, 1996. **24**(8): 1515–1524.

[42] Höhl M., Kurtz S., Ohlebusch E. Efficient multiple genome alignment. *Bioinformatics*, 2002. **18** (suppl_1): S312–S320.

[43] James K.W. BLAT–the BLAST-like alignment tool. *Genome Research*, 2002. **12**(4): 656–664.

[44] Homer N., Merriman B., Nelson S.F. BFAST: An alignment tool for large scale genome resequencing. *PloS One*, 2009. **4**(11).

[45] Brudno M., et al. LAGAN and Multi-LAGAN: Efficient tools for large-scale multiple alignment of genomic DNA. *Genome Research*, 2003. **13**(4): 721–731.

[46] Heath L.S., Ramakrishnan N. Problem Solving Handbook in Computational Biology and Bioinformatics. Springer Science & Business Media; 2010.

[47] Sievers F., Higgins D.G. Clustal omega. *Current Protocols in Bioinformatics*, 2014. **48**(1): 3.13.1–3.13.16.

[48] Sievers F., Barton G.J., Higgins D.G. Multiple sequence alignments. *Bioinformatics*, 2020 Feb 12. **227**: 227–250.

[49] McWilliam H., et al. Analysis tool web services from the EMBL-EBI. *Nucleic Acids Research*, 2013. **41** (W1): W597–W600.

[50] Lobo I. Basic local alignment search tool (BLAST). *Nature Education*, 2008. **1**(1): 1–9.

[51] Mount D.W. Using the basic local alignment search tool (BLAST). *Cold Spring Harbor Protocols*, 2007. **2007**(7): 17.

[52] Camacho C., et al. BLAST+: Architecture and applications. *BMC Bioinformatics*, 2009. **10**(1): 421.

[53] Wheeler D., Bhagwat M. BLAST Quickstart. In *Comparative Genomics*, Springer; 2007, 149–175.

[54] Madden T. The BLAST Sequence Analysis Tool. In The NCBI Handbook [Internet], 2nd Edn, National Center for Biotechnology Information (USA); 2013.

[55] Boratyn G.M., et al. Domain enhanced lookup time accelerated BLAST. *Biology Direct*, 2012. **7**: 12.

[56] Dereeper A., et al. BLAST-EXPLORER helps you building datasets for phylogenetic analysis. *BMC Evolutionary Biology*, 2010. **10**(1): 8.

[57] Ye J., et al. Primer-BLAST: A tool to design target-specific primers for polymerase chain reaction. *BMC Bioinformatics*, 2012. **13**: 134.

[58] Ye J., et al. IgBLAST: An immunoglobulin variable domain sequence analysis tool. *Nucleic Acids Research*, 2013. **41**(Web Server issue): W34–40.

[59] Thompson J.D., Higgins D.G., Gibson T.J. CLUSTAL W: Improving the sensitivity of progressive multiple sequence alignment through sequence weighting, position-specific gap penalties and weight matrix choice. *Nucleic Acids Research*, 1994. **22**(22): 4673–4680.

[60] Altschul S.F., et al. Basic local alignment search tool. *Journal of Molecular Biology*, 1990. **215**(3): 403–410.

[61] Finn R.D., Clements J., Eddy S.R. HMMER web server: Interactive sequence similarity searching. *Nucleic Acids Research*, 2011. **39**(suppl_2): W29–W37.

[62] Buchfink B., Reuter K., Drost H. Brief Communication Sensitive protein alignments at tree-of-life scale using DIAMOND. *Nature Methods*. https://doi.org/10.1038/s41592-021-01101-x.

[63] Kang C.-H., So J.-S. Antibiotic and heavy metal resistance in *Shewanella putrefaciens* strains isolated from shellfishes collected from West Sea, Korea. *Marine Pollution Bulletin*, 2016. **112**(1–2): 111–116.

[64] Oehmen C.S., Baxter D.J. ScalaBLAST 2.0: Rapid and robust BLAST calculations on multiprocessor systems. *Bioinformatics*, 2013. **29**(6): 797–798.

[65] Aßhauer K.P., et al. Tax4Fun: Predicting functional profiles from metagenomic 16S rRNA data. *Bioinformatics*, 2015. **31**(17): 2882–2884.

[66] Wheeler W., Gladstein D. MALIGN: A multiple sequence alignment program. *Journal of Heredity*, 1994. **85**(5): 417–418.

[67] Sadreyev R., Grishin N. COMPASS: A tool for comparison of multiple protein alignments with assessment of statistical significance. *Journal of Molecular Biology*, 2003. **326**(1): 317–336.

[68] Eddy, S. R., & Durbin, R. (1995). "RNA sequence analysis using covariance models." Nucleic Acids Research, **22**(11), 2079–2088. doi:10.1093/nar/22.11.2079

**Multiple choice questions**

Q1 What is the main purpose of Clustal W?
   a) Predicting protein–protein interactions
   b) Identifying conserved regions in multiple sequences
   c) Predicting RNA secondary structures
   d) Analyzing gene expression patterns

Q2 What type of sequence alignment does Clustal W perform?
   a) Pairwise alignment
   b) Multiple sequence alignment
   c) Global alignment
   d) Local alignment

Q3 Which algorithm is used by Clustal W for multiple sequence alignment?
   a) Smith–Waterman
   b) Needleman–Wunsch
   c) Dynamic Programming
   d) Clustal Omega

Q4 Which of the following best describes the purpose of T-Coffee?
   a) Identifying conserved regions in multiple sequences
   b) Predicting protein–protein interactions
   c) Identifying open reading frames in DNA sequences
   d) Visualizing phylogenetic trees

Q5 What does the "T" in T-Coffee signify?
   a) Time-efficient alignment
   b) Tree-based alignment
   c) Temperature-based alignment
   d) Tool-based alignment

Q6 Which algorithm is commonly used for performing multiple sequence alignment?
   a) Smith–Waterman
   b) Needleman–Wunsch
   c) Clustal Omega
   d) BLAST

Q7 Which of the following algorithms is specifically designed for aligning protein sequences?
   a) Clustal Omega
   b) MUSCLE
   c) T-Coffee
   d) MAFFT

Q8 Which of the following methods is commonly used for constructing phylogenetic trees based on molecular data?
   a) Maximum likelihood
   b) Punnett squares
   c) Mendelian inheritance
   d) Hardy–Weinberg equilibrium

Q9 Which software is widely used for phylogenetic analysis?
   a)   Microsoft Excel
   b)   Adobe Photoshop
   c)   MEGA (Molecular Evolutionary Genetics Analysis)
   d)   AutoCAD

Q10 In pairwise alignment, what does a positive score indicate?
   a)   A match between residues
   b)   A mismatch between residues
   c)   The presence of gaps in the alignment
   d)   The absence of gaps in the alignment

**Answers**
Q1   b)   Identifying conserved regions in multiple sequences
Q2   b)   Multiple sequence alignment
Q3   d)   Clustal Omega
Q4   a)   Identifying conserved regions in multiple sequences
Q5   b)   Tree-based alignment
Q6   c)   Clustal Omega
Q7   a)   Clustal Omega
Q8   a)   Maximum likelihood
Q9   c)   MEGA (Molecular Evolutionary Genetics Analysis)
Q10  a)   A match between residues

Ramesh C. Thakur*, Akshay Sharma, Renuka Sharma

# Chapter 5
# Recent advances in the discovery of drug molecules: trends, scope, and relevance

**Abstract:** The use of drugs for medicinal purposes has been practiced by humans since ancient times. In the early days, drugs were derived from animal sources, microbial sources, and mineral sources. As the pharmaceutical industry advanced, the synthesis of synthetic drugs became popular. However, speedy and low-cost drug discovery remains a challenge as drug development expenses are increasing due to longer and more expensive clinical trials. When a drug enters the body, it undergoes a journey from the point of entry to the site of action, which broadly includes the following phases: absorption, distribution, metabolism, and excretion (ADME). A potential drug needs to have favorable ADME properties. Natural products offer a vast source of leads as they provide a rich supply of therapeutically beneficial compounds. The search for new pharmacologically active agents through screening natural sources such as microbial fermentation and plant extracts has led to the discovery of many clinically beneficial drugs that play an essential role in the treatment of human diseases. In this chapter, we have described the importance of drugs, their sources and types, the journey of a drug from being identified as a lead, the criteria necessary for a lead to become an effective drug, and lastly, the role of natural products as potential leads in drug design for curing a variety of chronic illnesses.

**Keywords:** Natural products, drug design, partition coefficient, biological activity, drug lead

## 5.1 Introduction

In ancient times, health and diseases were considered a question of balance, with food and herbs classified by their ability to affect natural homeostasis. Before the eighteenth century, pharmacy remained a pseudoscience derived from traditional medicine. Mostly, plant extracts were used for the preparation of drugs. However, these drugs did not conform to the chemical definitions of the modern world. Later, in the nineteenth century, during the industrial revolution, drugs attracted industries

*Corresponding author: **Ramesh C. Thakur**, Department of Chemistry, Himachal Pradesh University, Summer Hill, Shimla 171005, Himachal Pradesh, India, e-mail: drthakurchem@gmail.com
**Akshay Sharma, Renuka Sharma**, Department of Chemistry, Himachal Pradesh University, Summer Hill, Shimla 171005, Himachal Pradesh, India

https://doi.org/10.1515/9783111568584-005

and became industrial items. The development of organic chemistry led to organic synthesis, and the growing pharmaceutical industry made drugs strategic items, mainly during colonial expeditions and for military purposes.

The French term "drogue-dried herbs" is where the word "drug" first appeared. Although it might be difficult to accurately classify medications, drugs may have several meanings depending on the circumstances, drug control legislation, medicine, and other factors. Drugs are defined by the World Health Organization (WHO) as any component used in a pharmaceutical product that modifies or alters pathological conditions or physiological systems for the benefit of the receiver. A chemical compound having a recognized chemical formula and structure that, when taken by an organism, induces physiological and psychological changes is also referred to as a drug. A pharmaceutical substance, often known as a medication or medicine, is used to treat, prevent, or diagnose a disease as well as to promote healthy living. Drugs are defined as "compounds that interact with a biological system to produce a biological response" in another definition. These criteria allow us to categorize numerous chemicals that, at first glance, we may not think of as medicines. For example:

– Morphine – interacts with the body to deliver comfort.
– The venom of a snake – comes in contact with the body and may cause death!
– Strychnine – comes into contact with the body and may even cause death!
– LSD – also reacts with the body to communicate unpleasant feelings.
– Coffee – interacts with the body to wake you up.
– Penicillin – binds to bacterial cells and destroys them.
– Sugar – interacts with the tongue to communicate flavor.

So, we can say that drugs have many effects on living organisms from small to large and from good to bad as well. We found that not all drugs provide us only beneficial effects; they can also be treated as potential poisons. In medicinal chemistry, there is a term known as the "therapeutic index," which indicates the safety index of a particular drug. It measures the effect based on its concentration and the beneficial effect of the drug at low doses compared to the harmful effect at high doses. A high therapeutic index means that there is a high level of safety between the beneficial doses and the toxicity. The prices for marijuana and alcohol are 1,000 and 10, respectively.

From different definitions of drugs, we found that a drug can be any compound that alters our biological responses when it encounters biological systems. They may be obtained by any method, whether natural or manmade. Before the twentieth century, medicines were synthesized from crude animal and mineral extracts, and in the twentieth century, we saw the rise of synthetic chemistry. In the process of finding and developing medicine, scientists began using several synthetic and semisynthetic artificial approaches. The development of novel, adaptable technologies, including automated separation techniques, high-performance testing, and combinatorial chemistry, has facilitated the identification and discovery of potentially medicinal chemicals. The medications we swallow, inject, and breathe in are typically complex medicinal substances.

A drug is often a mixture of chemical compounds created from raw materials, mostly plants, which have served as an essential source of unique molecules directly helpful as medicinal agents from the beginning of time. Drugs may be categorized as natural, synthetic, or semisynthetic depending on the sources from which they were derived.

## 5.2 Natural drugs

These items are made from natural materials, and the fact that they are widely available and affordable makes them accessible for rural people, meaning that their active ingredients are essential to the advancement of modern medicine. They are also an important source of a variety of bioactive substances, which may be used directly or as models in studies on medical ethnobotany, which examines the complex interaction between people and plants.

Natural product molecules have significant medical value beyond their pharmacological and chemotherapeutic effects because they may be used as models to create novel medications. For example, the morphine generated from the opium poppy has been used as a basic building block for the development of several drugs, such as analgesics like pethidine and pentazocine and the cough suppressant dextromethorphan. It remains very effective in treating extreme pain, especially in situations of terminal illness.

Untapped bioactive chemicals are widely distributed in nature and come from a variety of sources, including microorganisms, ants, frogs, worms, plants, animals, and marine species, to mention a few. These substances include, among others, atropine, the marine antiviral drug acyclovir, the antiprotozoal drug apicidin, ephedrine, caffeine, salicylic acid, digoxin, taxol, galantamine, vinblastine, vincristine, colchicine, quinine, artemisinin, etoposide, teniposide, and paclitaxel. Natural sources provide a consistent supply of outstanding clusters of bioactive lead chemicals for the creation of new medicines. There are thought to be 250,000 species of higher plants, many of which will become extinct over the next 100 years. The majority of the more than 20,000 species of higher plants used in traditional medicine have not recently been the subject of chemical or pharmacological research despite the fact that they are used in traditional medicine. It is well known that plants produce a wide variety of compounds with various chemical structures and pharmacological effects. The extensive range of compounds contained in plant extracts may be fractionated using tests for the binding of substances to receptors or enzymes, which are employed in modern pharmacology techniques. Innovative drugs may now be produced and isolated utilizing a combination of chemical, biological, and molecular biological techniques. Spectroscopy and chromatography-based separation and shape analysis techniques are quite sensitive. Such study offers a practical method for identifying model molecules for the development of novel therapies and the development of innovative medical treatments. Natural medicines may come from plant, animal, microbial, marine, mineral, or geographic sources.

## 5.2.1 Plant, animal, and microbial sources

Plant medications may be made from a variety of plant materials including the whole plant, particular plant parts, floral secretions, and exudates. Ergot, ephedra, and datura are a few whole plants utilized for medical reasons. Other plant parts, including senna leaves and pods, Digitalis leaves (which contain cardiotonic digoxin), Cinchona bark (the source of antimalarial quinine), opium tablets (containing analgesic morphine), Nux vomica seeds, Eserine seeds (containing anticholinesterase serine-physostigmine), and ginger rhizomes, are important sources of various medications used to treat a variety of illnesses. However, after being harvested, it is usual practice to dry most of these plant components, and they are seldom utilized in their raw form. Citrus peels like lemon and orange, as well as colchicum corm, are a few examples of products that are an exception to this rule.

Crude medications may be obtained via simple physical processes like drying or water extraction. For instance, opium is derived from the dried latex of poppy capsules, while black catechu is the dried aqueous extract of the woody parts of Acacia catechu. The dried leaves of many aloe species are used to make aloe juice. Substances used by doctors, pharmacists, or other healthcare providers are also included in pharmacognosy. Materials like gums, wax, gelatin, and agar are included in this category because they are utilized as pesticides, flavoring agents, sweeteners, or pharmacological components in vehicles. Furthermore, materials like kaolin and diatomite are used to clear turbid liquids, while materials like cotton, silk, jute, and nylon are used to make surgical bandages. Alkaloids, glycosides, oils, resins, gums, tannins, and many other bioactive chemicals have long been abundant in plants, which are also a great source of pharmacologically active molecules.

## 5.2.2 Animal sources

Drugs made from animal sources may be derived from a variety of materials such as whole animals, organs, glandular products (such as thyroid organ extracts), and liver tissues. Examples of these sources include the hirudin and heparin-producing European medical leech *Hirudo medicinalis* and the Mexican scientific leech *Hirudo manillensis*. Additionally, irritants like cantharides are made from insects like *Mylabris* sp. and *Lytta vesicatoria*, which are often known as Spanish fly or blistering beetles and are members of the Coleoptera family. From the secretions of lac bugs (*Laccifer lacca*), a resinous material known as lac or shellac is produced.

This class of medications also includes many organs and the products they produce. For example, the skin of the African clawed frog, *Xenopus laevis*, is used to make antimicrobial peptides, while the skin extracts of the Ecuadorian poison frog, *Ameerega bilinguis*, are used to make powerful analgesic substances like epibatidine. Insulin and hormones are produced by the pancreas of pigs and cows, pepsin is produced by cow stomachs, and thyroxin is produced by the thyroid gland. Additionally, liver tissues are used for the extraction of several compounds, including liver extract and vitamin B12.

## 5.2.3 Microbial sources

Only a few of the essential drugs derived from microorganisms are penicillin, which comes from the fungus *Penicillium notatum*; chloramphenicol, which comes from the bacterium *Streptomyces venezuelae*; griseofulvin, an antifungal drug; neomycin, which comes from the actinobacterium *Streptomyces griseus*; and streptomycin. Additionally, the aminoglycosides gentamicin and tobramycin are produced by *Micromonospora* sp. and *Streptomyces tenebrarius*, respectively.

## 5.2.4 Marine sources

Coral reefs, sponges, marine life, and microbes have all been shown to have a broad spectrum of highly bioactive chemical compounds with fascinating features, including anti-inflammatory, antiviral, and anticancer activities. For instance, Curacin A, a lipid component derived from the marine cyanobacterium *Lyngbya majuscula*, has potent antitumor activities. It is one of these substances. Other antitumor substances produced by the marine environment include eleutherobin, which is found in the coral *Eleutherobia* sp., discodermolide, which is derived from the Caribbean marine sponge *Discodermia dissoluta*, and bryostatins, which are derived from the colonial marine organism *Bugula neritina*, discovered off the coast of North Carolina. Dolostatins, which are derived from the small marine

## 5.2.5 Mineral (metallic and nonmetallic) sources

Pharmaceutical supplies may include minerals such as kaolin, chalk, diatomite, and others. In medicine, minerals and salts have significant medicinal advantages. For instance, magnesium sulfate ($MgSO_4$) is used as a laxative, magnesium trisilicate, aluminum hydroxide ($Al(OH)_3$), and sodium bicarbonate ($NaHCO_3$) are used as antacids to treat hyperacidity and peptic ulcers, and zinc oxide ointment is used for skin care, including sun protection and the treatment of eczema and wound healing. Solganal and auranofin, two gold salts, are used as anti-inflammatory treatments for rheumatoid arthritis. Talc, a hydrated magnesium silicate, and bentonite, an absorbent aluminum phyllosilicate clay, are also used in a variety of medical settings.

### 5.2.6 Geographical or habitat sources

We may learn more about the nation or area that produces the medicine by looking at the geographic supply or habitat. *Zingiber officinale*, which originated in southern China, has since spread to other parts of Asia, the Spice Islands (in the Indonesian province of Maluku), West Africa, and the Caribbean. It has also been bred with *Cannabis indica, Tamarindus indica, Strychnosnux vomica*, and *Plantago ispaghula* on the Indian subcontinent.

## 5.3 Synthetic drugs or designer drugs

These medications are created using ingredients that don't exist in nature. Instead, they are made from more straightforward molecular building blocks. Chemical synthesis, which reorganizes chemical constituents to shape a new molecule, is the process used to create synthetic pharmaceuticals. Utilizing cutting-edge phytochemical research methodologies and human expertise, the synthetic medication sources were created in laboratories. The vast majority of modern pharmaceuticals and chemical compounds used in research are created synthetically in chemical and pharmaceutical facilities. Sulfonamide, one of the first synthetic pharmaceuticals, was produced while the color prontosil was being made. Acetylsalicylic acid, often known as aspirin or ASA, oral diabetic medications, antihistamines, thiazide diuretics, chloroquine, chlorpromazine, common and local anesthetics, paracetamol, phenytoin, etc., are some further examples. The improved quality, purity, and cost of synthetically made tablets are often linked to high yields.

### 5.3.1 Semisynthetic

Semisynthetic drugs fall between the natural and synthetic spectrums. They are hybrids, and to improve their power, effectiveness, and/or lessen side effects, they typically include chemically enhancing components that may be found in a natural source. When natural sources may potentially provide impure substances or when the synthesis of medicines (complex molecules) may also be challenging, costly, or economically unviable, semisynthetic techniques may be utilized to generate therapeutics. The chemical structure of pharmaceuticals made from natural sources is still present in semisynthetic medications, but the nucleus has been changed. Derivatives of 6-aminopenicillanic acid and semisynthetic human insulin are two examples, produced by mixing naturally existing compounds with other substances in order to boost potency and efficacy while reducing side effects. For instance, morphine is used to make heroin, scopolamine is used to make Bromo scopolamine, and atropine is used to make homatropine. Derivatives of 6-aminopenicillanic acid and animal insulin that have been modified to mimic human insulin are examples of substances derived from animals. Additional examples include apomorphine, diacetyl morphine, ethinyl estradiol, homatropine, ampicillin, and methyl testosterone.

## 5.4 Drug content

When we talk about the constituents of a drug, we classify them as mentioned below.

### 5.4.1 Active ingredient

The substance present in a medicinal drug that has the preferred therapeutic impact is the active ingredient. In most cases, the active ingredient makes up a very small section of a tablet, pill, capsule, or suspension liquid.

### 5.4.2 Excipient: inactive ingredient

Inactive substances should include the active component if it is too tiny to be packaged, handled, or even seen in a single dose. Excipients are the term used to describe these inactive components in the pharmaceutical and pharmacological fields.

Simply put, the medication's active ingredients make it work, and the excipients give it structure so humans can handle it.

## 5.5 Drug categories

Medical professionals classify drugs according to how they affect the body. The effects that pills have on the body after use are classified by experts who specialize in the identification of medications. Additionally, certain drugs may make the body slow down, while others may make the body speed up. Stimulants, inhalants, cannabinoids, depressants, opioids, steroids, hallucinogens, and prescription drugs are only a few of the many drug classes.

The idea that these drugs (like stimulants and depressants) have opposite effects and may even out any variances is a pervasive myth. This is untrue, and using drugs like this may be fatal. Tragic outcomes might also happen if the body is overworked and the heart starts to beat erratically. These chemicals have the potential to impair a person's capacity to manage their central nervous system and drive safely.

There are several medication varieties and dosage options. While certain things are unlawful, others are not. Drug abuse and addiction may cause a variety of health problems, and in severe situations, they can even be deadly. Treatment for drug abuse is often utilized as a recovery method.

## 5.5.1 Stimulants

Stimulants speed up the neurological system of the body and provide the user with a sensation of energy. They are also known as "uppers" because of their ability to make you feel extremely awake. Depressants have the exact opposite impact as stimulants. The user is typically left with feelings of sickness and a loss of energy when the effects of a stimulant wear off. Continuous usage of these medications, if not taken as directed, can have very negative effects on the individual. Drug treatment facilities are frequently advised to address these medications' severe, horrible side effects and the impact they have on life.

**Types of drugs:**
1. Ritalin
2. Cylert
3. Amphetamines
4. Methamphetamines
5. Cocaine

## 5.5.2 Inhalants

Consumers who use inhalants, sniff or puff, immediately see results. Unfortunately, unexpected mental harm can also be a result of these immediate repercussions. When inhalants are consumed, the body loses oxygen efficiency, causing a rapid pulse. Other effects include issues with the liver, lungs, and kidneys, a compromised sense of smell, difficulty walking, and disorientation.

**Types of drugs include:**
1. Gasoline
2. Laughing gas
3. Aerosol sprays
4. Paint thinner
5. Glues

### 5.5.3 Cannabinoids

These medicines cause feelings of euphoria, disorientation, memory loss, anxiety, an increase in heart rate as well as stumbling and slow reaction times.

**Types of drugs include:**
1. Hashish
2. Marijuana

### 5.5.4 Depressants

Depressants reduce your body's central nervous system activity. These medicines are sometimes referred to as "downers" since they appear to calm the user and cause the body to slow down. Although sleepiness is commonly a side effect, depressants are available by prescription to ease stress and rage. Drug therapy is advised to stop drug misuse since the "relaxation" these drugs provide is not a good sensation for the body.

**Types of drugs:**
1. Alcohol
2. GHB (gamma-hydroxybutyrate)
3. Methaqualone
4. Flunitrazepam
5. Benzodiazepines
6. Barbiturates
7. Tranquilizers

### 5.5.5 Opioids and morphine derivatives

Opioids and morphine derivatives may cause breathing problems, drowsiness, nausea, feelings of euphoria, and pain relief.

**Types of drugs include:**
1. Opium
2. Codeine
3. Hydrocodone bitartrate, acetaminophen
4. Oxycodone HCL
5. Morphine
6. Heroin
7. Fentanyl and fentanyl analogues

## 5.5.6 Anabolic steroids

Steroids are used to increase strength, bulk up the muscles, and improve athletic performance. Steroid side effects include baldness, cysts, greasy skin, acne, heart attacks, strokes, and voice changes. Another typical negative effect of anabolic steroids is hostility.

**Types of drugs include:**
1. Oxandrin
2. Stanozol
3. Dianabol
4. Anadrol
5. Durabolin

## 5.5.7 Hallucinogens

Changing sensations when using hallucinogens is common. These substances alter the mind and cause things that aren't actually there to appear to be there. Hallucinogens affect the body's ability to regulate itself, impacting things like voice and movement that typically express animosity. Heart failure, an elevated heart rate, raised blood pressure, and hormonal changes are some more unpleasant side effects of these medications.

**Types of drugs include:**
1. LSD (lysergic acid diethylamide)
2. Mescaline
3. Magic mushrooms
4. Psilocybin

## 5.5.8 Prescription drugs

Prescription drugs may be very beneficial when used appropriately and in accordance with a competent physician's instructions. These drugs may be used to control a wide range of symptoms, treat medical conditions, and aid in surgery. Alternatively, misuse and abuse of prescription drugs may have severe negative effects.

Among the several drug classifications are:
1. Opioids, including oxycodone, codeine, and morphine.
2. Benzodiazepines and barbiturates are drugs that weaken the central nervous system.

Methylphenidate and dextroamphetamine are two stimulants.

## 5.6 Drug design

The human population of Earth is becoming older. Among the most critical health conditions that need the creation of innovative pharmaceuticals are several forms of cancer, viral infections, diabetes, and neurological problems. Despite this barrier, speedy and low-cost medication development is a goal that contrasts sharply with the state of drug research today. Depending on the therapeutic area, it might take 12–15 years and up to $1 billion USD to create a single drug. The current focus on chronic illnesses necessitates longer and more expensive scientific trials, elevated safety concerns brought on by catastrophic failures in the market, and more expensive research technologies are just a few of the factors causing the drug development process to become more expensive. Drug molecules often display therapeutic activity on certain targets at the cellular level when they bind to receptors, which change the cellular machinery as a consequence. Before a drug molecule may have its pharmacological (pharmacodynamic) impact on the organism via interaction with its target, it must first travel through the body to the site of drug action. Pharmacokinetics is the study of how a medication moves from its point of introduction to its point of effect. These four steps – absorption, distribution, metabolism, and excretion – can be used to generally characterize this process. The first difficulty with an oral medication is getting enough of it past the intestinal wall and into the bloodstream. It is transported to the liver after absorption, where a collection of hepatic microsomal enzymes is responsible for modifying it. Additionally, certain molecules could be digested, while others might be excreted via the bile. If a medicine molecule survives this first-pass metabolism, it will enter arterial circulation and be distributed to the body along with the target tissue. Once the medication has had the desired therapeutic effect, it must be progressively eliminated from the body to prevent bioaccumulation. A pharmaceutical should also not have any negative side effects, such as interfering with the effects of any other medicines the patient may be taking. A common strategy to produce such interference is enzyme induction, a procedure in which one medicine activates an enzyme, changing the metabolism of a second drug.

Therefore, it is not unexpected that successful drugs available today have certain physicochemical characteristics, even if pharmaceutical chemical structures may vary greatly based on the need for complementary interactions with different target receptors. A drug's pharmacokinetics is mostly impacted by these traits; hence it's critical to have good absorption, distribution, metabolism, and excretion (ADME) properties. Possibly the most extensively utilized study in this area was conducted by Lipinski and colleagues at Pfizer, who statistically reviewed over 2,200 medications from the World Drug Index. The so-called "rule of five" attributes, as stated by Lipinski, are a simple group of definable traits that were selected from medication candidates that placed in the 90th percentile and moved on to phase II clinical trials. It is a four-rule algorithm, and the name of the first rule comes from the fact that many of the cut-off values are multiples of five.

## 5.7 Lipinski's rule

To be labeled as drug-like, a candidate must contain fewer than 10 hydrogen bond acceptors, fewer than five hydrogen bond donors, less than 500 Da in molecular weight, and a partition coefficient log $P$ below five.

If two or more of the requirements are not satisfied, the "rule of five" is intended to draw attention to potential bioavailability issues. Rotatable bonds, polar surface area, log $D$, and counts of nitrogen and oxygen atoms are other variables that have been utilized to forecast favorable drug metabolism and pharmacokinetics results.

To find out whether a lead may create a potent medication, the four different components of Lipinski's rule are reviewed. Because they have an impact on how effectively a drug may be absorbed, distributed, metabolized, and excreted, these four qualities are essential.

## 5.8 Hydrophobicity of a drug

The hydrophobicity of a medication influences both its ability to traverse cell membranes and its ability to engage receptors. Pharmaceuticals interact with bodily compartments that range in physical characteristics from lipid-rich (hydrocarbon) environments to very water-soluble habitats. Pharmacology and pharmacokinetics are fundamentally shaped by the peculiar behavior of molecules in such settings and their distribution among them. Chemists have a thorough understanding of how molecules behave in immiscible organic solvents and water, which serves as the basis for analyzing this behavior in drug development. Hydrophobic substances are those that do not easily dissolve in water, whereas hydrophilic substances do. Hydrophobicity is often used in this sense despite the fact that the IUPAC defines it as "the tendency of non-polar groups or molecules to associate in an aqueous environment due to water's aversion to non-polar molecules."

Lipid (fat) chains are hydrocarbon-rich and are the complete antithesis of watery environments. When referring to compounds that prefer to live in such a setting, the word "lipophilic" is occasionally used in place of "hydrophobic." Lipophilicity is described by the IUPAC as "the attraction of a molecule or a portion of a molecule to a lipophilic environment." This quartet's etymological conclusion, "lipophobic," is only seldom utilized.

The partitioning of molecules between aqueous buffers and an immiscible solvent has long been recognized as important for pharmacological action, and various experimental combinations have been recorded. The Hansch group recommended the best applicable system for drug development in 1964: $n$-octan-1-ol and an aqueous buffer.

## 5.9 Role of partition coefficient in drug design

The partition coefficient, commonly denoted as log $P$ (log10 POW), is found by analyzing the equilibrium concentrations of a sample that has been evenly split between $n$-octan-1-ol and an aqueous buffer layer after complete agitation to reach equilibrium. The natural hydrophobicity or lipophilicity of a material is expressed as a constant known as log $P$. When a molecule lacks any ionizable sites, the partitioning analysis is simplified since the pH of the aqueous solution has no effect on the ratio of concentration between the two phases. However, for the bulk of pharmaceutical compounds, which typically include one or more ionizable sites, the pH has a significant impact on the distribution of neutral and charged forms across the phases.

Each chemical compound has a unique log $P$ value that shows how the nonionized structure of the molecule partitions. A buffer solution with a pH that is much lower (for acids) or higher (for bases) than the $pK_a$ of the molecule's ionizable sites may also be used to establish this constant. The log $P$ value for the fully ionized state is represented by an extra asymptote that is calculated under opposing circumstances (pH, $pK_a$ for acids, and pH, $pK_a$ for amines). To get this figure, extrapolation could be required, however. Notably, the distribution behavior changes, and the log $P$ may become fictional when a molecule has two or more ionizable sites because neutral species may no longer exist at certain pH values. The distribution coefficient, abbreviated as D, describes the distribution of all species of a molecule between the aqueous phase and an immiscible solvent at a certain pH and takes into consideration the existence of ionizable cores. To fully comprehend how the molecule behaves in the aqueous-solvent combination, this coefficient is determined.

## 5.10 Biological activity of a drug

The biological action of a medication and its hydrophobic properties may both be significantly impacted by substituent changes. Therefore, being able to forecast this statistically is crucial. However, as different substituents are added to the lead chemical, a variety of analogues with differing hydrophobicities and subsequently different $P$ values are created. It is possible to determine whether there is a correlation between the two qualities by graphing these $P$ values against the biological activity of these medications. In general, biological activity is stated as $1/C$, where $C$ is the medication concentration needed to produce a specific level of biological activity. Because more potent active medicines would achieve the indicated biological activity at a lower concentration, the reciprocal of the concentration ($1/C$) is employed. In order to create the diagram, log ($1/C$) vs. log $P$ are plotted. The use of logarithms permits the use of additional manageable numbers as the scale of numbers used to measure $C$ and $P$ often spans a variety of components frequently. In research where the range of the log $P$

values is limited to a small range (e.g., log $P$ = 1–4), a straight-line graph is obtained, displaying that there is a relationship between hydrophobicity and biological activity. Such a line would have the following equation:

$$\log 1/C = k1 \log P + k2 \tag{5.1}$$

For example, the binding of drugs to serum albumin is determined by their hydrophobicity, and a study of forty compounds resulted in the following equation:

$$\log 1/C = 0.75 \log P + 2.30 \tag{5.2}$$

According to the formula, the binding of serum albumin increases as the log $P$ value increases. In simpler terms, hydrophobic drugs exhibit stronger binding to serum albumin compared to hydrophilic ones. Determining the most effective dosage levels for a drug often depends significantly on its binding affinity to serum albumin. Drug dosages should primarily consider the concentration of unbound drug in the bloodstream, as drugs cannot interact with their receptors while bound to serum albumin. The mentioned formula enables us to assess the binding strength of drugs with similar structural characteristics to serum albumin and provides insight into their potential for receptor interactions. It is commonly recognized that increasing the hydrophobicity of a lead compound leads to enhanced biological activity despite considerations related to serum albumin binding. This underscores the importance of drugs overcoming hydrophobic barriers like cell membranes to reach their target sites. Even in situations where no barriers are encountered, such as in vitro research, drugs must still interact with target systems featuring hydrophobic binding sites, like enzymes or receptors. Increasing hydrophobicity, therefore, aids in penetrating these barriers or binding to the intended sites. It might be assumed that increasing the log $P$ value would lead to an infinite increase in biological activity, but this is not the case. Several factors contribute to this phenomenon. For instance, a drug may become excessively hydrophobic, rendering it only slightly soluble in aqueous solutions. Alternatively, it could become "trapped" in fat deposits and fail to reach its target site. Additionally, highly hydrophobic drugs often undergo extensive metabolism and subsequent elimination. The tendency for log $P$ values to plateau frequently results in the observation of a linear relationship between log $P$ and biological activity in many quantitative structure–activity relationship studies. For example, previous research on serum albumin binding was limited to compounds with log $P$ values ranging from 0.78 to 3.82. If this study were extended to include substances with exceptionally high log $P$ values, we would observe a different pattern, with biological activity increasing as log $P$ rises until it reaches a maximum value. The optimal partition coefficient for biological activity corresponds to the highest log $P$ value (log $P_0$). Beyond this point, an increase in log $P$ leads to a decrease in biological activity.

## 5.11 Natural products as leads

Natural products contain a rich reserve of valuable therapeutic substances. They offer a wide range of possibilities, ranging from inflexible, structurally constrained compounds to highly adaptable substances. The emergence of combinatorial chemistry technology led to high hopes that it would become the primary source of numerous innovative structures and drug candidates, often referred to as new chemical entities. These were expected to have straightforward intellectual property considerations. Interestingly, since the early 1980s, the influence of natural products on drug discovery across various therapeutic domains has waned. Consequently, natural products saw limited utilization in the pharmaceutical sector beyond that time, with only a handful of major pharmaceutical companies showing interest.

## 5.12 Natural products-based drug development

Through the investigation of natural sources, including microbial fermentations and plant extracts, the search for new pharmacologically active chemicals has led to the discovery of various medicinally significant medications that are crucial for treating human illnesses. According to recent studies, 60% of anticancer and anti-infective drugs that are either now on the market or in advanced clinical trials are derived from natural substances. Before their precise molecular targets were determined, several of these natural product-based medications, including substances like cyclosporine, paclitaxel, and camptothecin derivatives, were first discovered through typical in vitro assays like antibacterial, antifungal, antiviral, antiparasitic, or cytotoxic tests.

Of utmost significance, despite the pharmaceutical industry's escalated research and development endeavors, there is an immediate need to identify fresh, active chemical structures that can serve as the foundation for high-quality drug development across various therapeutic domains. For instance, within the realm of anti-infective treatments, the rapid emergence and widespread dissemination of microorganisms displaying resistance to both hospital-acquired and community-acquired antimicrobial agents pose a severe global threat to public health. Regrettably, there are presently no effective therapies available for addressing these resurgent infectious conditions. Similarly, the management of stable neoplastic diseases like breast, lung, colon, pancreatic, and liver cancers, as well as the imperative to create novel cancer chemotherapeutic agents to counter the growing issue of multidrug resistance in various tumor types undergoing chemotherapy, constitute substantial unmet clinical requirements in the field of oncology.

Based on the most notable accomplishments in natural product research over the past decade to 15 years, as succinctly presented in this assessment, natural products offer an unparalleled source of molecular diversity for drug exploration and development.

Undoubtedly, they complement emerging sources of molecules like combinatorial libraries. The primary challenge in this domain lies in enhancing the competitiveness of natural product searches by incorporating synthetic and combinatorial libraries. This is primarily due to the time-intensive process of isolating and characterizing active compounds from the complex mixtures typically found in natural product extracts, which can render natural product research less efficient when compared to the synthetic and combinatorial library approaches.

## 5.13 Conclusion and future recommendations

There is a serious reason for worry given the prevalence of both communicable and noncommunicable illnesses as well as the challenges in developing drugs with little or no adverse effects. Despite the availability of treatments for diseases including HIV/AIDS, malaria, hypertension, diabetes, and cancer, these illnesses continue to have a high death rate in many communities globally. The existing pharmaceutical research and development strategy, which is focused on "blockbuster" pharmaceuticals, is outmoded, and there is a desire for novel and unorthodox methods of drug discovery. Investigating natural sources for prospective treatments is a promising strategy since nature has already produced effective medication candidates. Natural substances like Taxol from the *Taxus brevifolia* tree, Vinblastine from *Catharanthus roseus*, and antimalarial medications like quinine from *Cinchona* spp. and Artemisinin from *Artemisia annua* have all shown promise in the treatment of these illnesses. Natural products confront extra challenges because of accessibility issues, sustainable acquisition issues, and intellectual property restrictions despite the fact that medication development often suffers from high attrition rates. However, improvements in analytical apparatus with very sensitive detectors have improved the detection of tiny molecules in biological systems, opening new opportunities for the development of more creative drugs to treat diseases and disorders. Unquestionably, the study and development of natural products play a crucial role in modern medication discovery in light of concerns about global public health.

**Conflict of interest:** The authors declare no conflicts of interest.

# References

[1]    Sung Y.K., Kim S.W. Recent advances in polymeric drug delivery systems. *Biomaterials Research*, 2020. **24**: 12.

[2]    Atanasov A.G., Zotchev S.B., Dirsch V.M., et al. Natural products in drug discovery: Advances and opportunities. *Nature Reviews Drug Discovery*, 2021. **20** 200–216

[3]    Li Chun-Qiang L.-H.-M.-Y.-H.-J. Recent Advances in the Synthetic Biology of Natural Drugs, 2021, Frontiers in Bioengineering and Biotechnology.

[4]    The Practice of Medicinal Chemistry edited by Camille Georges Wermuth. 4th Edn, 2015.

[5]    Alamgir A.N.M. Drugs: Their Natural, Synthetic, and Biosynthetic Sources. In Therapeutic Use of Medicinal Plants and Their Extracts, Vol. 1, Cham: Springer; 105–123, 2017.

[6]    Rang H.P., Dale M.M., Ritter J.M., Flower R.J., Henderson G. Rang & Dale's Pharmacology, 7th Edn, Edinburgh: Churchill Livingstone; 1, 2011. ISBN 978-0-7020-3471-8.

[7]    Drug. Dictionary.com Unabridged. Volume 1.1. Random House. 20 September 2007. Archived from the original on 14 September 2007 – via Dictionary.com.

[8]    An Introduction to Medicinal Chemistry GRAHAM L. PATRICK Department of Chemistry, Paisley University (6ed.) ISBN: 9780198749691.

[9]    Historical Background to Drug Discovery. UGA Center for Drug Discovery. [http://cdd.rx.uga.edu/index.php/history/]

[10]   Maehle A.H., Prüll C.R., Halliwell R.F. *Nature Reviews Drug Discovery*, 2002. **1**(8): 637–641.

[11]   Historical Background to Drug Discovery. UGA Center for Drug Discovery. [http://www.uga-cdd.org/background.php]

[12]   Pereira D.A. *Journal of Phytopharmacology*, 2007 Sep. **152**(1): 53–61. Epub 2007 Jul 2. PUBMED.

[13]   Pandeya S.N. Combinatorial Chemistry: A Novel Method in Drug Discovery and Its Application, 2005.

[14]   Lahlou M. The success of natural products in drug discovery. *Pharmacology and Pharmacy*, 2014. **4**: 17–31. Indian Journal of Chemistry, Feb 2005. **44B**: 335–348.

[15]   Phillipson J.D. *Transactions of the Royal Society of Tropical Medicine and Hygiene*, 1994, 88: 17–19.

[16]   https://www.pharmapproach.com/sources-of-drugs/

[17]   Proudfoot J.R.B. *Medicinal Chemistry Letters*, 2002. **12**: 1647–1650.

[18]   https://casapalmera.com/blog/top-8-drug-categories/

[19]   Smith R.N., et al. *Journal Pharmaceutical Sciences*, 1975. **64**: 599–606.

[20]   Tute MS. Lipophilicity: A History. In: Mannhold R., et al (ed), Methods and Principles in Medicinal Chemistry, New York: Wiley; 1996, 7–26.

[21]   Fujita T., Iwasha J., Hansch C. *Journal of the American Chemical Society*, 1964. **86**: 5175–5180.

[22]   Leo A., Hansch C., Elkins D. *Journal of Chemical Reviews*, 1971. **71**: 525–616.

[23]   Tan D.S., Foley M.A., Shair M.D., Schreiber S.L. *Journal of the American Chemical Society*, 1998. **120**: 8565–8566.

[24]   Rouhi A.M. *Chem Engineering News*, 2003. **81**: 104–107.

[25]   Newman D.J., Cragg G.M., Snader K.M. *Journal of Natural Products*, 2003. **66**: 1022–1037.

[26]   Newman D.J., Cragg G.M. *Journal of Natural Products*, 2004. **67**: 1216–1238.

[27]   Lam K.S. *Trends in Microbiology*, 2007. **15**(6): 279–289.

[28]   Newman D.J., Cragg G.M. *Journal of Natural Products*, 2007. **70**: 461–477.

[29]   Thomford N.E., Senthebane D.A., Rowe A., Munro D., Seele P., Maroyi A., Dzobo K. *International Journal of Molecular Sciences*, 2018. **19**(6): 1578.

[30]   Cragg G.M., Newman D.J., Snader K.M. *Journaal of Natural Products*, 1997. **60**(1): 52–60.

[31]   Shu Y.Z. *Journal of Naturalproducts*, 1998. **61**(8): 1053–1071.

[32]   Newman D.J. *Journal Medical Chemistry*, 2008. **51**(9): 2589–2599.

**Multiple-choice questions**

Q1 What were the earliest sources of drugs used for medicinal purposes?
   a) Synthetic chemicals
   b) Animal, microbial, and mineral sources
   c) Plants and herbs
   d) Geological formations

Q2 Which phase of a drug's journey in the body involves its absorption, distribution, metabolism, and excretion?
   a) Manufacturing
   b) Clinical trials
   c) ADME
   d) Licensing

Q3 Why is the rapid and cost-effective discovery of new drugs challenging?
   a) Lack of research funding
   b) Shortage of pharmaceutical companies
   c) Increasing expenses in scientific trials
   d) Regulatory hurdles

Q4 What is the primary role of natural products in drug discovery?
   a) Acting as excipients
   b) Providing a source of therapeutically beneficial compounds
   c) Replacing synthetic drugs
   d) Enhancing drug distribution

Q5 Which of the following is not a common source of natural drugs?
   a) Marine organisms
   b) Animal sources
   c) Metallic minerals
   d) Microbial fermentation

Q6 What is the significance of Lipinski's rule in drug design?
   a) It determines the color of the drug.
   b) It assesses the hydrophobicity of a drug.
   c) It predicts a drug's biological activity.
   d) It analyzes the drug's taste.

Q7 Which drug category includes substances like caffeine and amphetamines?
   a) Stimulants
   b) Depressants
   c) Hallucinogens
   d) Cannabinoids

Q8 What is the role of partition coefficient in drug design?
   a) It measures the drug's solubility in water.
   b) It assesses the drug's stability.
   c) It predicts a drug's biological activity.
   d) It determines the drug's color.

Q9    Which category of drugs includes substances like heroin and codeine?
  a)   Stimulants
  b)   Opioids & Morphine Derivatives
  c)   Hallucinogens
  d)   Inhalants

Q10   What do semisynthetic drugs involve?
  a)   Drugs produced entirely in a laboratory
  b)   A combination of natural and synthetic elements
  c)   Drugs derived solely from animal sources
  d)   Drugs made from minerals

Q11   Which factor is not considered during drug design?
  a)   Biological activity of a drug
  b)   Partition coefficient
  c)   Drug's color
  d)   Hydrophobicity of a drug

Q12   Which source of drugs is derived from plant extracts?
  a)   Microbial sources
  b)   Animal sources
  c)   Geographical or Habitat sources
  d)   Synthetic drugs

Q13   Excipients in drug formulation are primarily
  a)   Active ingredients
  b)   Inactive ingredients
  c)   Natural products
  d)   Designer drugs

Q14   What type of drugs are used to induce a state of calmness and relaxation?
  a)   Stimulants
  b)   Depressants
  c)   Cannabinoids
  d)   Opioids & Morphine Derivatives

Q15   What is the role of natural products in drug design?
  a)   They are essential excipients.
  b)   They replace synthetic drugs entirely.
  c)   They provide a potential source of leads.
  d)   They regulate drug distribution.

Q16   Which rule assesses a drug's hydrophobicity?
  a)   Lipinski's rule
  b)   Partition coefficient rule
  c)   Biological activity rule
  d)   Color rule

**Q17** What is the primary focus of the chapter's discussion on drug categories?
a) Drug manufacturing processes
b) Drug classification based on chemical structure
c) Drug distribution methods
d) Drug marketing strategies

**Q18** What is the main topic of discussion in the chapter?
a) Drug discovery through natural products
b) Synthetic drug development
c) Clinical trials of new drugs
d) Drug pricing and affordability

**Q19** Which type of drug is typically prescribed by a healthcare professional?
a) Designer drugs
b) Opioids & Morphine Derivatives
c) Natural products
d) Over-the-counter drugs

**Q20** In drug discovery, what does ADME stand for?
a) Active Drug Manufacturing Environment
b) Absorption, Distribution, Metabolism, and Excretion
c) Advanced Drug Management Expertise
d) Analytical Drug Monitoring and Evaluation

---

**! Answers**

Q1  b
Q2  c
Q3  c
Q4  b
Q5  c
Q6  c
Q7  a
Q8  c
Q9  b
Q10 b
Q11 c
Q12 c
Q13 b
Q14 b
Q15 c
Q16 a
Q17 b
Q18 a
Q19 b
Q20 b

Meenakshi Rajpoot, Varruchi Sharma, Shagun Gupta, Ankur Kaushal, Anupam Sharma, Vandana Sharma, J. K. Sharma, Anil Panwar, Seema Ramniwas, Damanjeet Kaur, Anil K. Sharma*

# Chapter 6
# Computer-aided drug design and drug discovery

**Abstract:** One of the most important processes in the pharmaceutical business is drug discovery, and computational drug discovery is a significant technique for speeding up and minimizing the cost of the development process. Virtual screening, homology modeling, molecular docking, molecular dynamics simulations, and pharmacophore modeling are examples of computational technologies that can greatly reduce the time and cost of drug development. Large libraries of compounds are being screened and synthesized in a short amount of time due to the rapid development of combinatorial chemistry and high-throughput screening techniques, consequently speeding up the drug discovery process. Furthermore, as our understanding of three-dimensional biological structures has evolved, so have our computational capabilities, allowing us to employ computer-assisted techniques effectively at various stages of the drug development process. This chapter provides an overview of these critical computational methodologies, platforms, and their effective application in the creation of new drugs. Here, we discuss structure-based, ligand-based, and sequence-based drug discovery methods.

**Keywords:** Drug discovery, computer-aided drug design, virtual screening, molecular docking, Molecular simulations

*Corresponding author: Anil K. Sharma, Department of Biotechnology, Amity University Punjab, Sector 82 A, IT City Rd, Block D, Mohali 140306, Punjab, India, email: anibiotech18@gmail.com
Meenakshi Rajpoot, Department of Biotechnology, DAV College (Lahore), Ambala City, Ambala, Haryana, India
Varruchi Sharma, Damanjeet Kaur, Department of Biotechnology, Sri Guru Gobind Singh College, Sector 26, Chandigarh 160019, India
Shagun Gupta, Ankur Kaushal, Department of Bio-sciences and Technology, MMEC, Maharishi Markandeshwar (Deemed to be University), Mullana, Ambala 133207, Haryana, India
Anupam Sharma, Department of Physics, Guru Kashi University Talwandi Sabo, Bathinda-151302
Vandana Sharma, J. K. Sharma, Department of Physics, MMEC, Maharishi Markandeshwar (Deemed to be University), Mullana, Ambala 133207, Haryana, India
Anil Panwar, Department Department of Bioinformatics and Computational Biology, CCS Haryana Agricultural University, Hisar 125004, Haryana, India
Seema Ramniwas, University Centre for Research and Development, University Institute of Biotechnology, Chandigarh University, Gharuan, Mohali, Punjab, India

https://doi.org/10.1515/9783111568584-006

## 6.1 Introduction

For many years, drug discovery relied on experimental techniques, and despite numerous efforts to improve efficiency, this continued to be a costly and lengthy process [1]. A typical drug research process takes about 10–14 years and costs between 0.8 and USD 1.0 billion [2]. High prices, time-consuming procedures, high risk levels, and uncertain results are the key hurdles in innovative drug development. As a result, cost-effective drug design approaches such as computer-aided drug design and molecular docking became prominent. Proper usage of computer-based approaches has reduced the obstacles in the process and hastened novel drug development [3]. The combined application of modern medicinal chemistry methods is used to determine the pharmacokinetic and pharmacodynamic properties and QSAR of drug candidates with their target receptor [4, 5]. Implementing these techniques not only reduces the number of studies and pre-clinical trials on animals but also allows for more efficient management of large amounts of data and more exact test findings [6, 7]. CADD may also be used in conjunction with experimental techniques to elucidate drug-resistance mechanisms, identify novel antibiotic targets, and develop new antibiotics [8].

Computer-aided drug design, or CADD, refers to computational tools used for storage, management, data analysis, and molecular modeling [9]. It combines computer programs for lead design, software to analyze potential lead candidates, and database development for the study of chemical interactions [10]. A general representation of CADD is summarized in Figure 6.1. CADD approaches are mainly classified into three categories: structure-based drug design (SBDD), ligand-based drug design (LBDD), and sequence-based methods. SBDD mainly relies on the knowledge of the structure of target proteins obtained through X-ray crystallography, NMR data, or homology modeling.

However, the nonavailability of 3D structures leads to the LBDD approaches where QSAR, pharmacophore modeling, molecular field analysis, and 2D or 3D similarity techniques are used to provide in-depth knowledge of interactions between a drug molecule and target receptor [2, 11] and build predictive models suitable for lead discovery and optimization. Sequence-based approaches are basically bioinformatics tools that are used to analyze and compare multiple biological sequences when there is no information available about the target structure and the ligand [2, 12].

## 6.2 Structure-based drug design

A comprehensive understanding of the disease pathway and its related mechanisms is essential for drug design as it helps in the selection of an appropriate drug target. Therefore, some programs like GeneGo, KEGG, BioCyc, and MetaCyc are employed to construct pathway maps of diseases [13] to improve and hasten the drug development process and minimize the failure rates [14]. Structure-based drug design (SBDD) is

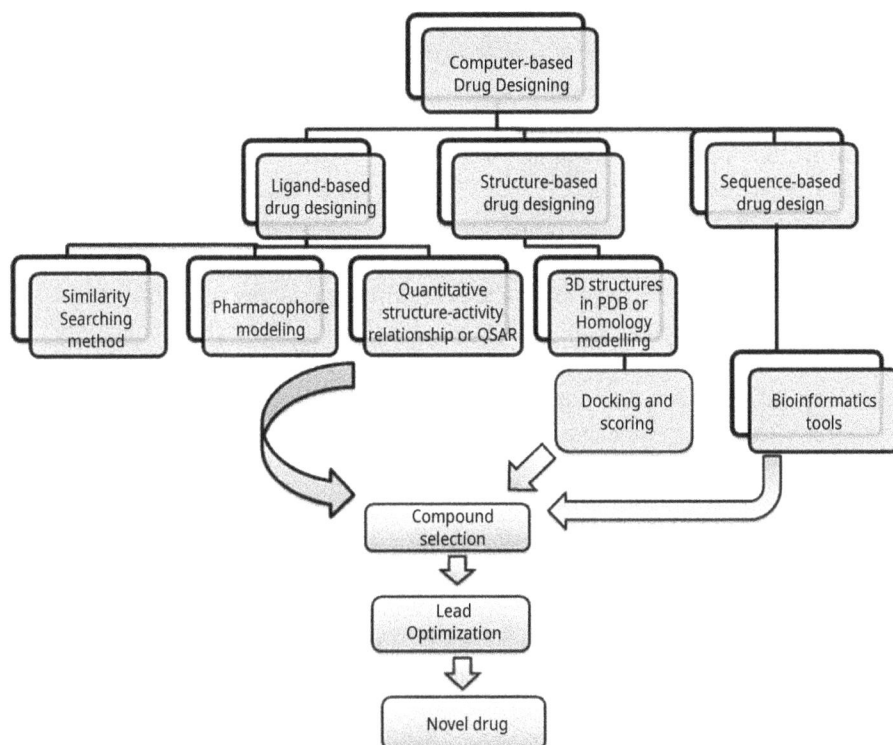

**Figure 6.1:** General representation of computer-aided drug designing (CADD).

mainly dependent on the availability of three-dimensional structures of target proteins as this helps in identifying their binding sites. It deals with the structure of the proteins along with molecular-level information about the disease; therefore, it is an efficient, specific, and rapid method for lead discovery and optimization [15, 16]. Apart from X-ray diffraction, nuclear magnetic resonance, and homology modeling, computational techniques such as virtual screening, molecular docking, and molecular dynamics simulations are also used in SBDD. All these techniques help in the binding energy analysis, drug–protein interactions, and observation of conformational changes that occur during molecular docking. Recent developments in the software industry have led to the massive availability of tools and software for these processes, and there is no doubt that automation of these procedures has shortened the time-span of SBDD [15, 17, 18]. Several drugs such as the HIV-1-inhibiting drug, thymidylate synthase inhibitor, raltitrexed, amprenavir, and the antibiotic norfloxacin are available in the market, identified by the SBDD process [16].

## 6.2.1 Protein structure determination and hot spot prediction

The availability of complete genome sequences results in a multitude of possible targets, and diverse techniques such as systems biology, probabilistic models, clustering, and drug affinity reactions aid in target identification. The initial stage in this procedure is structural identification and confirmation, which is often accomplished using X-ray crystallography and NMR. The resolved structures are available at the RCSB Protein Data Bank (PDB) [19]. A recent resolution revolution in cryo-electron microscopy has led to the determination of more and more structures at nearly atomic-level resolution [20]. If the experimentally determined structures of target proteins are not known, then computer-based techniques such as ab initio modeling, threading, and homology modeling are used for predicting the structures [13, 19, 21]. Homology or comparative modeling is the most significant and fastest method for structure prediction used in rational drug designing as well as for studying protein–protein interactions and site-directed mutagenesis. The 3D structure of an unknown protein can be built if its sequence similarity is above 30% with their homologous protein templates [13]. Various online servers (e.g., Swiss-Model, Modeller, etc.) enable the prediction of 3D structures. Homology models are produced by employing homologous structural regions of the sequences with a high degree of similarity, and these model structures could be useful in drug discovery [22].

Once the model is built, the next step is the determination of the active site for predicting protein–ligand interactions, post-docking dynamics, binding free energies, and the formation of hydrogen bonds [21]. If there is no data available regarding the binding site, a number of web-based tools such as DoGSite Scorer, CASTp, NSiteMatch, MetaPocket, and Q-SiteFinder [23–27] are available, which provide knowledge about the active sites. These methods give results by considering the interaction energy and van der Waals (vdW) forces for predicting the binding site and identifying the specific regions that interact significantly with the important functional groups of the ligand [27]. After this step, the binding pocket's volume is also evaluated with the help of some tools like Epock, Transient Pockets in Proteins (TRAPP), and POVME [28–30].

## 6.2.2 Virtual screening

Virtual screening, or vHTS, an alternative approach to high-throughput screening, is used to computationally screen huge chemical libraries and search for compounds complementary to the target receptors. This screening is accomplished by molecular docking, in which ligands are selected based on their binding affinities against the therapeutic target [1, 31]. VS is classified into two main categories: structure-based (SBVS) and ligand-based (LBVS). SBVS is useful when the structure of the target protein is known, and then the target protein is docked with large chemical libraries similar to drugs using computer algorithms. Scoring functions are used to evaluate the binding energies of the docked complexes along with experimental assays to validate the binding process; therefore, it is a crucial step. However, LBVS applies predictive models of compound activity based on available data and helps in separating inactive compounds from the active ones, which is useful in determining highly active scaffolds [18, 32]. Several inhibitors and antagonists have been explored with the aid of computer-based screening methods, and one such successful effort was the development of the drug candidate SC12267 for arthritis treatment [33]. Moreover, VS gives ten times higher hit-rates compared to empirical screening procedures, due to which it is equally important as high-throughput screening [34, 35].

## 6.2.3 Docking and scoring functions

Docking is basically a virtual simulation of molecular interactions, and it is a major step in SBDD as it helps in analyzing the interactions and binding affinity of protein–ligand complexes [21, 36–37]. It determines the conformation and binding of ligands at the active site with accuracy. A molecular phenomenon like the ligand-binding pose and intermolecular interactions for the stability of docked complexes can be studied with the aid of docking [16]. Ligand-binding conformation relies on a large conformational space which defines possible binding poses and the correlation of binding energy with each conformation. Therefore, it is necessary to perform multiple iterations until the minimum energy state is obtained and assess ligand-binding by scoring functions [4, 38]. Various algorithms like genetic algorithms, evolutionary algorithms, Monte Carlo algorithm, simulated annealing algorithms, and the fast Fourier transform approach are used for effective parameter space searching [39].

In molecular docking, it is important to determine the binding affinity to see how well a drug binds to the target receptor, which is performed by scoring. Scoring functions are used to score and evaluate protein–ligand complexes [20]. Scoring functions are mainly of three types: force-field analysis, empirical, and knowledge-based scoring functions. Force-field analysis measures the actual molecular forces present between the ligand and protein by improving van der Waals, hydrophobic, and electrostatic energy to each force field. Some of the force-field analysis functions are GoldScore, CHARMM, Amber, and OPLS [40, 41]. The empirical scoring function, such as TS-Chemscore, optimizes van der Waals interaction energies, hydrogen-bond energy, hydrophobicity, desolvation energy, electrostatic forces, and entropy to calculate the interactions between the atoms interacting with each other and alteration in the solvent-accessible surface area [42, 43], is the fastest of all the scoring functions. The third one, i.e., knowledge-based scoring function, works by using geometrical data from empirical structures from databases such as PDB [44]. A list of frequently used docking software along with their scoring functions and applications is summarized in Table 6.1.

**Table 6.1:** Some frequently used molecular docking software.

| Software | Scoring functions | Main applications |
|---|---|---|
| DOCK Version 6.6 | Force field | Predict binding modes of ligand–protein complexes; search databases of ligands for compounds that inhibit enzyme activity<br>Examine possible binding orientations of protein–protein and protein–DNA complexes |
| AutoDock Version4.2.5 | Force field, empirical | X-ray crystallography; SBDD; lead optimization; virtual screening; combinatorial library design; protein–protein docking; chemical mechanism studies |
| Glide | Empirical | Virtual screening<br>Accurate binding mode prediction<br>Fully prepared databases of purchasable compounds from Enamine, MilliporeSigma, and MolPort |
| FleX Version 2.1.3 | Empirical | Binding mode prediction<br>Virtual high-throughput screening (vHTS) |
| Affinity | Force field | Predict the binding model of the ligand and target<br>Optionally, model flexibility in the target macromolecule |
| Fred Version 2013 | Empirical | Perform a systematic and non-stochastic examination of all possible protein–ligand poses<br>Provide a detailed scoring analysis<br>Fast virtual screening programs |
| SLIDE Version 3.4 | Force field, empirical | Handle large binding-site templates and use multistage indexing to identify feasible<br>Subsets of template points for ligand docking<br>Screen 100,000 compounds within a few days and return a ranked list of sterically feasible<br>Ligand candidates, ranked by complementarity to the protein's binding site |

# 6.3 Ligand-based drug designing

Ligand-based drug design (LBDD) doesn't search for small molecule libraries but depends on the understanding of known small molecules that could bind to the target protein of interest [45]. This approach relies on statistical methods to find the relationship between ligand activity and structural information [13]. Some of the LBDD methods are pharmacophore modeling, molecular similarity, and QSAR (quantitative structure–activity relationship) modeling [12]. While molecular similarity approaches use molecular fingerprints of known small molecules that bind to the target receptor to search for ligands with the same fingerprints by screening chemical libraries, pharmacophore modeling relies on the common structural characteristics of the ligand molecules that bind to a similar target protein for screening. QSAR uses the structural relationship between the ligands that bind to a target to study the effects of associated biological activity [46–48].

## 6.3.1 Similarity searching method

The main approach behind the similarity-based method is that it selects novel compounds that have similar chemical and physical properties to known drug molecules for the target protein. This method is based on the principle that structurally similar molecules have similar binding characteristics but does not consider information related to the activities of known ligands for the therapeutic target [49]. It provides an option when there is no information regarding the 3D structure of the target protein or it cannot be predicted. Therefore, it is mostly employed to screen novel ligands having distinct biological activities to improve drug pharmacokinetic properties such as ADMET, i.e., Absorption, Distribution, Metabolism, Excretion, Toxicity [50]. Similarity searching programs are mainly classified into 2D and 3D similarity. The former is efficient for fast profiling of neighboring compounds, but sometimes it provides different hits for the same queries, offering close structural analogs instead of novel scaffolds. The latter one considers multiple aspects of pharmacophores, molecular shapes, and molecular fields, and it can be used to achieve lead hopping so that novel compounds can be identified [51–52]. For the G-protein-coupled target GPR30, a particular agonist that activates GPR30 has been developed using similarity searches [20].

## 6.3.2 Pharmacophore modeling

A pharmacophore is a molecular organization that describes the essential features behind the biological activity of a ligand molecule and explains how structurally distinct ligand molecules can bind to a single binding site of a receptor [52]. When a drug interacts with a target protein, this gives rise to a geometrically as well as energetically matching active conformation of the ligand with the target macromolecule. Some medicinal chemists discovered that different functional groups present on drug molecules pose distinct effects on their activity. Alterations in some groups highly affect the drug–target interactions, whereas changes in other groups have a slight influence [53]. Also, it is noted that molecules having similar activity possess some of the same features [45]. As either the target structure or known ligand molecules can be used to construct the models, pharmacophore-based screening can be seen as the crossing between SBDD and LBDD approaches and is termed receptor-based pharmacophore and ligand-based pharmacophore [54].

Pharmacophore models are built with the help of structural features of active ligands that bind to a particular target when there is limited or no information available about the structure of the drug target [40]. However, when 3D structure data of the target protein is known, this information is used to build pharmacophore models [55]. The most effective models are those that use chemical features like acidic-basic residues and H-bond acceptors and donors [56]. Pharmacophore modeling has been used in the virtual screening of ligand molecules in large databases [57]. DISCO, GASP, and Catalyst are some of the programs developed to identify and generate pharmacophore models [58].

## 6.3.3 Quantitative structure–activity relationship

Quantitative structure–activity relationship (QSAR) is a method that demonstrates a relationship between a ligand's structure and the biological or chemical activity of a target using mathematical models [59–60]. A QSAR model describes the effect of a specific property on the molecule's activity, unlike a pharmacophore model that encodes only the vital characteristics of an active ligand molecule [52]. When the receptor structure is not available, QSAR can be a precise and practical approach for drug design. 3D-QSAR explores the 3D perception of biologically active molecules, correctly shows the energy changes and interaction patterns between bioactive molecules and target receptors, and uncovers the drug-receiving mechanism based on the 3D structural features of both ligands and targets. Physicochemical parameters, as well as 3D structural parameters of a sequence of drugs, are appropriate for establishing the quantitative relationship. Finally, the structures of novel compounds are predicted and optimized [45].

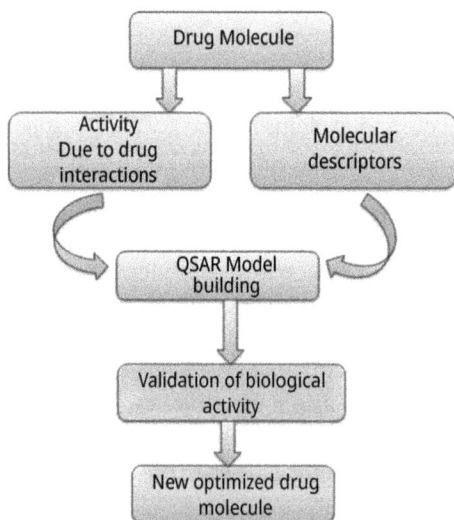

**Figure 6.2:** Schematic diagram showing the steps involved in QSAR.

The process of QSAR model construction involves the collection of ligands and their activities (Figure 6.2). Before choosing the mathematical modeling method, molecular descriptors are calculated. The ligand data is then used to construct the QSAR models, followed by testing using both internal and external validation strategies [52, 61].

## 6.4 Sequence-based drug design

Three-dimensional structures of many proteins are not available, and there are no known ligands for several proteins. In this case, both structure-based approaches and ligand-based methods do not work in conducting drug design and development. Therefore, a sequence-based method for ligand–protein interactions was developed that took the primary sequence of proteins and the structural characteristics of ligand molecules by using a support vector machine (SVM) approach [62]. This model was trained with 15,000 ligand–protein interactions between a total of 626 proteins and more than 10,000 active ligand compounds from the Binding Database [63]. Nine new active compounds were discovered by employing the sequence of the target [64].

In sequence-based approaches, a given drug molecule is screened against thousands of proteins constituting a human proteome. These methods employ information about the drug as well as proteins to determine drug–protein interactions. Based on the availability of the 3D structure of the target protein, these methods can be divided into two groups: one is protein structure-based and the other is a similarity-based approach [64–66]. The first one requires the three-dimensional structure of target proteins to generate and run predictive models [67–69]. A recent study reveals that approximately 20–30% of human proteins have three-dimensional structures, considering both experimentally produced and computationally predicted structures [70–72]. The similarity-based approach considers an internal database of known drug–protein interactions and a predictive model that generates presumed interactions employing the database [73–76].

## 6.5 Role of molecular dynamics simulations in drug discovery

Molecular dynamics simulation is a theoretical approach used to inspect the conformations and dynamic behaviors of biomolecules [77, 78]. It is based on classical Newtonian mechanics and applies empirical molecular mechanics force fields. On the basis of different accuracy levels, all-atom, united-atom, coarse-grained, temporal, and spatial scales simulation can be used to ease the drug discovery process [79, 80]. MD simulation has been proven as a complement to wet-lab procedures to fill the gap between knowledge of three-dimensional structures of target proteins and their potential inhibitors [81].

The 3D models resolved by NMR, X-ray crystallography, and homology modeling provide valuable information about the macromolecular structure; however, they are static. When a ligand approaches its target in solution, it does not encounter a single and static structure but a macromolecule in constant movement. Protein motions are limited only in a few cases when the ligand may fit into a fixed binding pocket. Hence, MD simulations have crucial roles in drug discovery [79]. MD simulations are mainly used to identify the potential drug binding sites on therapeutic targets, calculate binding free energy, and provide insight into the mechanism of action of drugs and inhibitors, etc. [82, 83]. Yan et al. conducted extensive simulations and solvent-accessible surface area calculations to examine the interactions in the binding cavity for the PRMT1-diamidine complex and provided a direction toward designing more potent and specific inhibitors [84–86]. A similar MD study was also reported to propose binding patterns of known PRMT1 inhibitors [11, 87]. This method had many successful applications, including a cancer-relevant MDM2/MDMx-p53 interaction or HIV integrase [88].

## 6.6 Success stories of computational drug discovery approaches

Computational approaches have proven to play a vital role in modern drug discovery. The applications of CADD methods in drug discovery, especially anticancer, have proven to be great not only in terms of investment but also in resources and time. Numerous novel anticancer drugs such as Abiraterone, Gefitinib, Erlotinib, Lapatinib, Sorafenib, and Crizotinib have been successfully discovered using CADD, and all are FDA approved [89–93]. Anticancer drug research is rapidly progressing, and computational and Artificial Intelligence methods are producing promising results. A potent inhibitor was successfully identified for 5-lipoxygenase using a machine learning method developed by employing physicochemical and pharmacophore characteristics [94, 95]. Luminespib, a drug in phase I of clinical trials, showed good results in patients with ALK rearrangements [96]. Moreover, it exhibited strong antitumor activity in lung adenocarcinomas [97, 98] and is also included as a constituent in anticancer combination therapies, presently at distinct stages of clinical trials [51, 99, 100].

Recently, a group of researchers developed novel chromone-based copper(II) antitumor inhibitors by employing a combination of X-ray crystal structures and molecular docking [45, 101]. Computer-based approaches have also gained success in discovering novel antibiotics and their targets. A recent example is the novel antibiotic target, protein heme-oxygenase, which is involved in the heme metabolism of bacteria and is needed to access iron [102–104]. In a study, CADD techniques helped in identifying the inhibitors of bacterial heme oxygenases taken from bacterial species Pseudomonas aeruginosa and Neisseria meningitidis, and confirmed the potential role of heme-oxygenases as new antimicrobial targets [105, 106]. With in-silico database screening, a new group of non-β-lactam antibiotics, the oxadiazoles, was discovered, and it can inhibit penicillin-binding protein 2a of methicillin-resistant Staphylococcus aureus, the major cause of infections in hospitals [8, 106]. The success stories of drug discovery by SBDS are listed in Table 6.2.

**Table 6.2:** The successful cases of drug discovery by SBDD methods.

| Drug | Drug target | Target disease | Technique |
|---|---|---|---|
| Raltitrexed | Thymidylate synthase | Human immunodeficiency virus (HIV) | SBDD |
| Amprenavir | Antiretroviral protease | HIV | Protein modeling and molecular dynamics (MD) |
| Isoniazid | InhA | Tuberculosis | Structure-based virtual screening (SBVS) and pharmacophore modeling |
| Pim-1 Kinase Inhibitors | Pim-1 kinase | Cancer | Hierarchical multistage virtual screening |
| Epalrestat | Aldose reductase | Diabetic neuropathy | MD and SBS |
| Flurbiprofen | Cyclooxygenase-2 | Rheumatoid arthritis, osteoarthritis | Molecular docking |
| STX-0,119 | STAT3 | Lymphoma | SBVS |
| Norfloxacin | Topoisomerase II, IV | Urinary tract infection | SBVS |
| Dorzolamide | Carbonic anhydrase | Glaucoma, cystoid macular edema | Fragment-based screening |

## 6.7 Conclusion and future perspectives

Computational approaches play a pivotal role in drug screening and design. Successful implementation of these methods leads to the identification of biologically active compounds without being partial toward known ligand molecules. Methods such as docking and molecular dynamics simulations help in resolving multiple types of mechanisms involved in complex drug–target interactions. Virtual screening can be employed to search huge chemical libraries for lead compounds. Significant development of new software is continuously benefiting the field of pharmacokinetics and pharmacodynamics, which ultimately not only improves the process of drug discovery but also solves the problem of the high cost of these processes. These methods have proven to be constructive in multiple tasks such as protein interaction network analysis, target prediction, binding site prediction, screening of small molecules, and so on, which could significantly facilitate the designing of anticancer drugs. Furthermore, the availability of several computer-based methods has reduced the time as well as the number of experiments on animals. Drug discovery still faces a lot of challenges like upgrading the virtual screening methods, improving computational chemogenomic studies, enhancing the quality and number of computational databases, improving the multi-target drug structure, developing informatics toxicology, and collaborating with other related study fields for better results. No single software package is capable of working with all types of proteins and ligands, and despite ongoing improvements and developments, a consistent technique is yet to be developed.

Machine learning, a subfield of artificial intelligence, is revolutionizing CADD, which could enhance the efficiency and accuracy of the process. These integrated computer-based methods will speed up drug development and propose effective therapies with novel action mechanisms that could be used in various complex biological systems. Experts believe that computational technologies, along with artificial intelligence, have the capability to permanently transform the pharmaceutical industry and the process of drug discovery. In the future, AI and medicinal chemists can work together, as the medicinal expert would help in analyzing huge datasets and the AI expert would train machines, set algorithms, and optimize the analyzed data for a faster and more accurate drug development process. Multidisciplinary efforts should be employed to generate diverse datasets, which further help to thoroughly use the chemical space available in databases. With the current development of techniques, there is a promising future for computational methods in the discovery of many more therapeutics.

# References

[1]    Phatak S.S., Stephan C.C., Cavasotto C.N. High-throughput and in silico screenings in drug discovery. *Expert Opinion on Drug Discovery*, 2009. **4**: 947–959. doi: 10.1517/17460440903190961.
[2]    Ou-Yang S.S., Lu J.Y., Kong X.Q., Liang Z.J., Luo C., Jiang H. Computational drug discovery. *Acta Pharmacologica Sinica*, 2012. **33**(9): 1131–1140. https://doi.org/10.1038/aps.2012.109.
[3]    Jamkhande P.G., Ghante M.H., Ajgunde B.R. Software based approaches for drug designing and development: A systematic review on commonly used software and its applications. *Bulletin of Faculty of Pharmacy. Cairo University*, 2017. **55**: 203–210.
[4]    Ferreira L.G., Santos R.N., Oliva G., Andricopulo A.D. Molecular docking and structure-based drug design strategies. *Molecules*, 2015. **20**: (2015) 13384–13421.
[5]    Sliwoski G., Kothiwale S., Meiler J., Lowe E.W. Computational methods in drug discovery, Pharmacol. *Review*, 2014. **66**: 334–395.
[6]    Hodos R.A., Kidd B.A., Shameer K., Readhead B.P., Dudley J.T. In silico methods for drug repurposing and pharmacology, computational approaches to drug repurposing and pharmacology. *Wires Systems Biology and Medicine*, 2016. **186**.

[7]   Liu T., Lu D., Zhang H., Zheng M., Yang H. Applying high-performance computing in drug discovery and molecular simulation, Natl. *Scientific Review*, 2016. **3**: 49–63.

[8]   Yu W., MacKerell A.D. Jr. Computer-Aided Drug Design Methods. *Meth Mol Biol (Clifton, N J )*, 2017. **1520**: 85–106. https://doi.org/10.1007/978-1-4939-6634-9_5.

[9]   Mohs R.C., Greig N.H. Drug discovery and development: Role of basic biological research. *Alzheimer's & Dementia (New York, N. Y.)*, 2017. **3**(4): 651–657. https://doi.org/10.1016/j.trci.2017.10.005.

[10]  Song C.M., Lim S.J., Tong J.C. Recent advances in computer-aided drug design. *Briefings in Bioinformatics*, 2009. **10**: 579–591.

[11]  Lu W., Zhang R., Jiang H., Zhang H., Luo C. Computer-Aided Drug Design in Epigenetics. *Frontiers in Chemistry*, 2018. **6**: 57. Published 2018 Mar 12doi: 10.3389/fchem.2018.00057.

[12]  Acharya C., Coop A., Polli J.E., Mackerell A.D. Jr. Recent advances in ligand-based drug design: Relevance and utility of the conformationally sampled pharmacophore approach. Computer-Aided Drug Design, 2011. **7**: 10–22.

[13]  Huang H.J., Yu H.W., et al. Current developments of computer-aided drug design. *Journal of the Taiwan Institute of Chemical Engineers*, 2010. **Volume 41**(Issue6): Pages 623–635. ISSN 1876-1070 https://doi.org/10.1016/j.jtice.2010.03.017.

[14]  Kaur D., Mathew S., Nair C.G.S., et al. Structure based drug discovery for designing leads for the non-toxic metabolic targets in multi drug resistant Mycobacterium tuberculosis. *J. Translational Medicine*, 2017. **15**(1): 261. doi: 10.1186/s12967-017-1363-9.

[15]  Lionta E., Spyrou G., Vassilatis D.K., Cournia Z. Structure-based virtual screening for drug discovery: Principles, applications and recent advances. *Current Topics in Medicinal Chemistry*, 2014. **14**: 1923–1938.

[16]  Batool M., Ahmad B., Choi S. A Structure-Based Drug Discovery Paradigm (2009) *Int. International Journal of Molecular Sciences*, **20**(11): 2783. Published 2019 Jun 6.doi: 10.3390/ijms20112783.

[17]  Searls D.B. Data integration: Challenges for drug discovery. *Nature Reviews Drug Discovery*, 2005. **4**: 45–58.

[18]  Kalyaanamoorthy S., Chen Y.P. Structure-based drug design to augment hit discovery. *Drug Discovery Today*, 2011. **16**: 831–839. www.rcsb.org/.

[19]  Leelananda S.P., Lindert S. Computational methods in drug discovery. *Beilstein Journal of Organic Chemistry*, 2016. **12**: 2694–2718. https://doi.org/10.3762/bjoc.12.267.

[20]  Wang X., Song K., Li L., Chen L. Structure-based drug design strategies and challenges. . *Current Topics in Medicinal Chemistry*, 2018. **18**(12): 998–1006. doi:. 10.2174/1568026618666180813152921.

[21]  Rajpoot M., Bhattacharyya R., Gupta G., Sharma A.K. "2. Drug Designing in Novel Drug Discovery: Trends, Scope and Relevance". In Chemical Drug Design, Berlin, Boston: De Gruyter, 2016. doi: https://doi.org/10.1515/9783110368826-004.

[22]  Volkamer A., Kuhn D., Rippmann F., Rarey M. DoGSiteScorer:A web server for automatic binding site prediction, analysis and druggability assessment. *Bioinformatics*, 2012. **28**(15): 2074–2075.

[23]  Sahu A., Patra P.K., Yadav M.K., Varma M. Identification and characterization of ErbB4 kinase inhibitors for effective breast cancer therapy. *Journal of Receptors and Signal Transduction*, 2017. **37**(5): 470–480.

[24]  Sun J., Chen K. NSiteMatch: Prediction of binding sites of nucleotides by identifying the structure similarity of local surface patches. *Computational and Mathematical Methods in Medicine*, 2017. **2017**: 5471607.

[25]  Huang B. MetaPocket: A meta approach to improve protein ligand binding site prediction. *Omics*, 2009. **13**(4): 325–330.

[26]  Laurie A., Jackson R. Q-SiteFinder: An energy-based method for the prediction of protein-ligand binding sites. *Bioinformatics*, 2005. **21**: 1908–1916.

[27]   Laurent B., Chavent M., Cragnolini T., et al. Epock: Rapid analysis of protein pocket dynamics. *Bioinformatics*, 2015. **31**(9): 1478–1480.

[28]   Stank A., Kokh D.B., Horn M., Sizikova E., et al. TRAPP webserver: Predicting protein binding site flexibility and detecting transient binding pockets. *Nucleic Acids Research*, 2017. **45**(W1): W325–W330.

[29]   Wagner J.R., Sorensen J., Hensley N., Wong C., Zhu C., Perison T., Amaro R.E. POVME 3.0: Software for mapping binding pocket flexibility. *Journal of Chemical Theory and Computation*, 2017. **13**(9): 4584–4592.

[30]   Shoichet B.K. Virtual screening of chemical libraries. *Nature*, 2004. **432**: 862–865.

[31]   Pedretti A., Mazzolari A., Gervasoni S., Vistoli G. Rescoring and linearly combining: A highly e_ective consensus strategy for virtual screening campaigns. *International Journal of Molecular Sciences*, 2019. **20**: 2060.

[32]   Coumar M.S., et al. Structure-based drug design of novel aurora kinase A inhibitors: Structural basis for potency and specificity. *Journal of Medicinal Chemistry*, 2009. **52**: 1050–1062.

[33]   Doman T.N., et al. Molecular docking and high-throughput screening for novel inhibitors of protein tyrosine phosphatase-1B. *Journal of Medicinal Chemistry*, 2002. **45**: 2213–2221.

[34]   Schneider G. Virtual screening: An endless staircase? *Nature Reviews Drug Discovery*, 2010. **9**: 273–276. https://www.rcsb.org/structure/6eha.

[35]   Jug G., Anderluh M., Tomasic T. Comparative evaluation of several docking tools for docking small molecule ligands to DCSIGN. *Journal of Molecular Modeling*, 2015. **21**(6): 164.

[36]   Kapetanovic I.M. Computer-aided drug discovery and development (caddd): Insilico-chemico-biological approach. *Chemico-Biological Interactions*, 2008. **171**: 165–176.

[37]   Huang S.Y., et al. Scoring functions and their evaluation methods for protein–ligand docking: Recent advances and future directions. *Physical Chemistry Chemical Physics*, 2010. **12**: 12899–12908.

[38]   Tropsha A. Cheminformatics meets molecular mechanics: A combined application of knowledge-based pose scoring and physical force field-based hit scoring functions improves the accuracy of structure-based virtual screening. *Journal of Chemical Information and Modeling*, 2012. **52**(1): 16–28.

[39]   Debroise T., Shakhnovich E.I., Cheron N. A hybrid knowledgebased and empirical scoring function for protein-ligand interaction:SMoG2016. *Journal of Chemical Information and Modeling*, 2017. **57**(3): 584–593.

[40]   Li G.B., Yang L.L., Wang W.J., Li L.L., Yang S.Y. ID-Score: A new empirical scoring function based on a comprehensive set of descriptors related to protein-ligand interactions. *Journal of Chemical Information and Modeling*, 2013. **53**(3): 592–600.

[41]   Eldridge M.D., Murray C.W., Auton T.R., Paolini G.V., Mee R.P. Empirical scoring functions: I. The development of a fast empirical scoring function to estimate the binding affinity of ligands in receptor complexes. *Journal of Computer-Aided Molecular Design*, 1997. **11**(5): 425–445.

[42]   Lizunov A.Y., Gonchar A.L., Zaitseva N.I., Zosimov V.V. Accounting for intraligand interactions in flexible ligand docking with a pmf-based scoring function. *Journal of Chemical Information and Modeling*, 2015. **55**(10): 2121–2137.

[43]   Lin X., Li X., Lin X. A review on applications of computational methods in drug screening and design. *Molecules*, 2020. *(Basel, Switzerland)* **25**(6): 1375. https://doi.org/10.3390/molecules25061375.

[44]   Vogt M., Bajorath J. In Bajorath J. Ed. Chemoinformatics and Computational Chemical Biology, Totowa, NJ: Humana Press, 2011, 2011 159–173.

[45]   Yang S.Y. Pharmacophore modeling and applications in drug discovery: Challenges and recent advances. *Drug. Discov Today*, 2010. **15**(11–12): 444–450. doi:. 10.1016/j.drudis.2010.03.013.

[46]   Verma J., Khedkar V.M., Coutinho E.C. 3D-QSAR in drug design–a review. *Curr. Topics in Medicinal Chemistry*, 2010. **10**(1): 95–115. doi:. 10.2174/156802610790232260.

[47]   Klopmand G. Concepts and applications of molecular similarity, by Mark A. Johnson and Gerald M. Maggiora, eds. *Journal of computational chemistry*, 1992. 13, 539–540. doi: 10.1002/jcc.540130415.

[48]  Cui W., Aouidate A., Wang S., Yu Q., Li Y., Yuan S. Discovering Anti-Cancer Drugs *via* Computational. *Methods. Frontiers in Pharmacology*, 2020. **11**: 733. https://doi.org/10.3389/fphar.2020.00733.

[49]  Liu X.F., Jiang H.L., Li H.L. SHAFTS: A hybrid approach for 3D molecular similarity calculation. 1. Method and assessment of virtual screening. *Journal of Chemical Information & Modeling*, 2011. **51**: 2372–2385.

[50]  Lee C.H., Huang H.C., Juan H.F. Reviewing ligand-based rational drug design: The search for an ATP synthase inhibitor. *International Journal of Molecular Sciences*, 2011. **12**(8): 5304–5318. https://doi.org/10.3390/ijms12085304.

[51]  Seidel T., Bryant S.D., Ibis G., Poli G., Langer T. (2017)3D pharmacophore modeling techniques in computer-aided molecular design using LigandScout. *Tutor Chemistry*, **281**: 279–309.

[52]  Wilson G.M., Muftuoglu Y. Computational Strategies in Cancer Drug Discovery, Advances in Cancer Management. In Mohan R. Ed, ISBN: 978–953-307-870-0, InTech, 2012, Available from. http://www.intechopen.com/books/advances-in-cancer-management/computational-strategies-in-cancerdrug-discovery.

[53]  Yang S.Y. Pharmacophore modeling and applications in drug discovery: Challenges and recent advances. *Drug Discovery Today: Technologies*, 2010. **15**: 444–450. doi: 10.1016/j.drudis.2010.03.013.

[54]  Lin S.K. Pharmacophore Perception, Development and Use in Drug Design. In Güner O.F. Ed, Molecules, vol. 5, 987–989, 2000. doi: 10.3390/50700987.

[55]  Langer T., Krovat E.M. Chemical feature-based pharmacophores and virtual library screening for discovery of new leads. *Curr. Opin Drug Discov Devel*, 2003. **6**(3): 370–376.

[56]  Patel Y., Gillet V.J., Bravi G., Leach A.R.J. A comparison of the pharmacophore identification programs: Catalyst, DISCO and GASP. *Computer-aided Molecular Design*, 2002. **16**: 653–681. doi: 10.1023/A:1021954728347.

[57]  Kubinyi H. Success Stories of Computer-Aided Design. In Ekins S., Wang B., Eds, Computer Applications in Pharmaceutical Research and Development, Hoboken, NJ, USA: Wiley-Interscience, 377–424.

[58]  Esposito E.X., Hopfinger A.J., Madura J.D. Methods for applying the quantitative structure activity relationship paradigm. *Methods in Molecular Biology*, 2004. **275**: 131–214.

[59]  Myint K.Z., Xie X.Q. Recent advances in fragment-based QSAR and multi-dimensional QSAR methods. *International Journal of Molecular Sciences*, 2010. **11**: 3846–3866.

[60]  Wang F., Liu D.X., Wang H.Y., Luo C., Zheng M.Y., Liu H., *et al.* Computational screening for active compounds targeting protein sequences: Methodology and experimental validation. *Journal of Chemical Information & Modeling*, 2011. **51**: 2821–2828.

[61]  Liu T.Q., Lin Y.M., Wen X., Jorissen R.N., Gilson M.K. BindingDB: A web-accessible database of experimentally determined protein-ligand binding affinities. *Nucleic Acids Research*, 2007. **35**: D198–201.

[62]  Ding H., Takigawa I., Mamitsuka H., Zhu S. Similarity-based machine learning methods for predicting drug-target interactions: A brief review. *Brief Bioinform*, 2014. **15**(5): 734–747.

[63]  Schomburg K.T., Rarey M. What is the potential of structure-based target prediction methods? *Future Medicinal Chemistry*, 2014. **6**(18): 1987–1989.

[64]  Somody J.C., MacKinnon S.S., Windemuth A. (2017) Structural coverage of the proteome for pharmaceutical applications. *Drug Discovery Today.*, **22**(12):1792–1799. doi:10.1016/j.drudis.2017.08.004

[65]  Brylinski M., Feinstein W.P. eFindSite: Improved prediction of ligand binding sites in protein models using meta-threading, machine learning and auxiliary ligands. *Journal of Computer-Aided Molecular Design*, 2013. **27**(6): 551–567.

[66]  Feinstein W.P., Brylinski M. eFindSite: Enhanced fingerprint-based virtual screening against predicted ligand binding sites in protein models. *Molecular Informatics*, 2014. **33**(2): 135–150.

[67]   Litfin T., Zhou Y., Yang Y. SPOT-ligand 2: Improving structure-based virtual screening by bindinghomology search on an expanded structural template library. *Bioinformatics*, 2017. **33**(8): 1238–1240.

[68]   Mizianty M.J., Fan X., Yan J., Chalmers E., Woloschuk C., Joachimiak A., Kurgan L. Covering complete proteomes with X-ray structures: A current snapshot. *Acta Crystallographica. Section D, Biological Crystallography*, 2014. **70**(Pt11): 2781–2793.

[69]   Liu T., Altman R.B. Relating essential proteins to drug side-effects using canonical component analysis: a structure-based approach. *Journal of Chemical Information and Modeling*, 2015. **55**(7): 1483–1494.

[70]   Zhang Q.C., Petrey D., Deng L., et al. Structure-based prediction of protein-protein interactions on a genome-wide scale. *Nature*, 2012. **490**(7421): 556–560.

[71]   Mitchell J.B. The relationship between the sequence identities of alpha helical proteins in the PDB and the molecular similarities of their ligands. *Journal of Chemical Information and Computer Sciences*, 2001. **41**(6): 1617–1622.

[72]   Schuffenhauer A., Floersheim P., Acklin P., Jacoby E. Similarity metrics for ligands reflecting the similarity of the target proteins. *Journal of Chemical Information and Computer Sciences*, 2003. **43**(2): 391–405.

[73]   Klabunde T. Chemogenomic approaches to drug discovery: Similar receptors bind similar ligands. *British Journal of Pharmacology*, 2007. **152**(1): 5–7.

[74]   Wang C., Kurgan L. Survey of Similarity-Based Prediction of Drug-protein Interactions. *Current Medicinal Chemistry*, 2019. **26**: 1. https://doi.org/10.2174/0929867326666190808154841.

[75]   Lindorff-Larsen K., Piana S., Dror R.O., Shaw D.E. How fast folding proteins fold. *Science*, 2011. **334**: 517–520.

[76]   Okumura H., Higashi M., Yoshida Y., Sato H., Akiyama R. Theoretical approaches for dynamical ordering of biomolecular systems. *Biochimica Et Biophysica Acta*, 2018. **1862**: 212–228. doi: 10.1016/j. bbagen.2017.10.001.

[77]   Durrant J.D., McCammon J.A. Molecular dynamics simulations and drug discovery. *BMC Biology*, 2011. **9**: 71.

[78]   Borhani D.W., Shaw D.E. The future of molecular dynamics simulations in drug discovery. *Journal of Computer-Aided Molecular Design*, 2012. **26**: 15–26.

[79]   Le L. Incorporating Molecular Dynamics Simulations into Rational Drug Design: A Case Study on Influenza A Neuraminidases, IntechOpen: Bioinformatics, Horacio Pérez-Sánchez, 2012, DOI 10.5772/52642.

[80]   Wang Y., Lupala C.S., Liu H., Lin X. Identification of drug binding sites and action mechanisms with molecular dynamics simulations. *Current Topics in Medicinal Chemistry*, 2018. **18**: 2268–2277.

[81]   Hou T., Wang J., Li Y., Wang W. Assessing the performance of the MM/PBSA and MM/GBSA methods. 1.The accuracy of binding free energy calculations based on molecular dynamics simulations. *Journal of Chemical Information and Modeling*, 2011. **51**: 69–82.

[82]   Yan L., Yan C., Qian K., Su H., Kofsky-Wofford S.A., Lee W.C., et al. Diamidine compounds for selective inhibition of protein arginine methyltransferase 1. *Journal of Medicinal Chemistry*, 2014. **57**: 2611–2622. doi: 10.1021/jm401884z.

[83]   Zhang J., Qian K., Yan C., He M., Jassim B.A., Ivanov I., et al. (2017a).Discovery of decamidine as a new and potent PRMT1 inhibitor. *Medchemcomm*, **8**: 440–444. doi 10.1039/C6MD00573J.

[84]   Zhang W.Y., Lu W.C., Jiang H., Lv Z.B., Xie Y.Q., Lian F.L., et al. Discovery of alkyl bis(oxy)dibenzimidamide derivatives as novel protein arginine methyltransferase 1 (PRMT1) inhibitors. *Chemical Biology & Drug Design*, 2017b. **90**: 1260–1270. doi: 10.1111/cbdd.13047.

[85]   Yang H., Ouyang Y., Ma H., Cong H., Zhuang C., Lok W.T., et al. Design and synthesis of novel PRMT1 inhibitors and investigation of their binding preferences using molecular modelling. *Bioorganic & Medicinal Chemistry Letters*, 2017. **27**: 4635–4642. doi: 10.1016/j.bmcl.2017.09.016.

[86]  Śledź P., Caflisch A. Protein structure-based drug design: From docking to molecular dynamics. *Current Opinion in Structural Biology*, 2017. 2018(**48**): 93–102. doi: 10.1016/j.sbi.2017.10.010.

[87]  Jarman M., Barrie S.E., Llera J.M. The 16,17-double bond is needed for irreversible inhibition of human cytochrome P450(17 alpha) by abiraterone (17-(3-pyridyl)androsta-5,16-dien-3 beta-ol) and related steroidal inhibitors. *Journal of Medicinal Chemistry*, 1998. **41**: 5375–5381. doi: 10.1021/jm981017j.

[88]  Muhsin M., Graham J., Kirkpatrick P. Fresh from the pipeline -Gefitinib. *Nature Reviews. Drug Discovery*, 2003. **2**: 515–516. doi: 10.1038/nrd1136.

[89]  Grunwald V., Hidalgo M. Development of the epidermal growth factor receptor inhibitor Tarceva (TM) (OSI-774). *New Trends in Cancer for the 21st Century*, 2003. 235–246. doi: 10.1007/978-1-4615-0081-0_19.

[90]  Wood E.R., Truesdale A.T., Mcdonald O.B., Yuan D., Hassell A., Dickerson S.H., et al. A unique structure for epidermal growth factor receptor bound to GW572016 (Lapatinib): Relationships among protein conformation, inhibitor off-rate, and receptor activity in tumor cells. *Cancer Research*, 2004. **64**: 6652–6659. doi: 10.1158/0008-5472.CAN-04-1168.

[91]  Wilhelm S., Carter C., Lynch M., Lowinger T., Dumas J., Smith R.A., et al. Discovery and development of sorafenib: A multikinase inhibitor for treating cancer. *Nature Reviews. Drug Discovery*, 2006. **5**: 835–844. doi: 10.1038/nrd2130.

[92]  Reker D., Rodrigues T., Schneider P., Schneider G. Identifying the macromolecular targets of de novo-designed chemical entities through selforganizing map consensus. *Proceedings of the National Academy of Sciences of the United States of America*, 2014. **111**: 4067–4072. doi: 10.1073/pnas.1320001111.

[93]  Rodrigues T., Werner M., Roth J., Da Cruz E.H.G., Marques M.C., Akkapeddi P., et al. Machine intelligence decrypts -lapachone as an allosteric 5-lipoxygenase inhibitor. *Chemical Science*, 2018. **9**: 6. doi: 10.1039/c8sc02634c.

[94]  Felip E., Barlesi F., Besse B., Chu Q., Gandhi L., Kim S.-W., et al. Phase 2 Study of the HSP-90 Inhibitor AUY922 in previously treated and molecularly defined patients with advanced non-small cell lung cancer. *Journal of Thoracic Oncology*, 2018. **13**: 576–584. doi: 10.1016/j.jtho.2017.11.131.

[95]  Jorge S.E., Lucena-Araujo A.R., Yasuda H., Piotrowska Z., Oxnard G.R., Rangachari D., et al. EGFR exon 20 insertion mutations display sensitivity to hsp90 inhibition in preclinical models and lung adenocarcinomas. *Clinical Cancer Research*, 2018. **24**: 6548–6555. doi: 10.1158/1078-0432.CCR-18-1541.

[96]  Piotrowska Z., Costa D.B., Oxnard G.R., Huberman M., Gainor J.F., Lennes I.T., et al. Activity of the Hsp90 inhibitor luminespib among non-smallcell lung cancers harboring EGFR exon 20 insertions. *Annals of Oncology*, 2018. **29**: 2092–2097. doi: 10.1093/annonc/mdy336.

[97]  Garcia-Carbonero R., Carnero A., Paz-Ares L. Inhibition of HSP90 molecular chaperones: Moving into the clinic. *Lancet Oncology*, 2013. **14**: E358–E369. doi: 10.1016/S1470-2045(13)70169-4.

[98]  Rong B., Yang S. Molecular mechanism and targeted therapy of Hsp90 involved in lung cancer: New discoveries and developments (Review). *International Journal of Oncology*, 2018. **52**: 321–336. doi: 10.3892/ijo.2017.4214.

[99]  Arjmand F., Afsan Z., Roisnel T. Design, synthesis and characterization of novel chromone based-copper(ii) antitumor agents with N, N-donor ligands: Comparative DNA/RNA binding profile and cytotoxicity. *RSC Advances*, 2018. **8**: 37375–37390.

[100]  Ballester P.J. Machine learning for molecular modelling in drug design. *Biomolecules*, 2019. **9**: 216.

[101]  Rydzewski J., Nowak W. Machine learning based dimensionality reduction facilitates ligand di_usion paths assessment: A case of cytochrome P450cam. *Journal of Chemical Theory and Computation*, 2016. **12**: 2110–2120.

[102]  Karplus M. Development of multiscale models for complex chemical systems: From H+ H2 to biomolecules (Nobel lecture). *Angewandte Chemie*, 2014. **53**: 9992–10005.

[103]  Jorgensen W.L. The many roles of computation in drug discovery. *Science*, 2004. **303**: 1813–1818.

[104]  De Vivo M., Masetti M., Bottegoni G., Cavalli A. Role of molecular dynamics and related methods in drug discovery. *Journal of Medicinal Chemistry*, 2016. **59**: 4035–4061.

[105]  Abel R., Wang L., Harder E.D., Berne B., Friesner R.A. Advancing drug discovery through enhanced free energy calculations. *Accounts of Chemical Research*, 2017. **50**: 1625–1632.

[106]  Mak K.K., Pichika M.R. Artificial intelligence in drug development: Present status and future prospects. *Drug Discovery Today*, 2019. **24**(3): 773–780. doi:. 10.1016/j.drudis.2018.11.014.

---

### Multiple choice questions

Q1  What is the main role of molecular docking in CADD (computer-aided drug design)?
- a)  Predicting the three-dimensional structure of proteins
- b)  Identifying potential binding sites on target proteins
- c)  Determining the optimal dosage of a drug
- d)  Evaluating drug safety in clinical trials

Q2  What do you understand by the term virtual screening in the reference to computer-aided drug design?
- a)  Conducting clinical trials using computer simulations
- b)  Screening large chemical libraries to identify potential drug candidates
- c)  Generating physical prototypes of drug molecules
- d)  Monitoring the effects of drugs on patients in real-time

Q3  Which of the following is used to simulate the interaction between a drug molecule and its target protein?
- a)  Molecular dynamics
- b)  Genetic algorithms
- c)  Support vector machines
- d)  Artificial neural networks

Q4  What does the term "pharmacophore" refer to?
- a)  The optimal pH for drug administration
- b)  The three-dimensional arrangement of functional groups essential for biological activity
- c)  The process of synthesizing pharmaceuticals in the laboratory
- d)  The regulatory agency responsible for approving new drugs

Q5  Which software tool is commonly used for molecular modeling and visualization in the drug design process?
- a)  Microsoft Excel
- b)  AutoCAD
- c)  Schrödinger Suite
- d)  Adobe Photoshop

Q6  Name the technique which is commonly used to identify pharmacophores?
- a)  Molecular docking
- b)  Molecular dynamics simulations
- c)  QSAR modeling
- d)  Pharmacophore modeling

Q7 Define the purspose of validating a pharmacophore model?
 a) To confirm its effectiveness in predicting drug activity
 b) To optimize the synthesis of pharmaceuticals
 c) To determine the optimal route of drug administration
 d) To identify potential side effects of drugs

Q8 What type of data is typically used as input for QSAR modeling?
 a) Biological assays
 b) Structural information of proteins
 c) Chemical properties of molecules
 d) Gene expression data

Q9 Which one is best describing the goal of Quantitative Structure–Activity Relationship (QSAR) modeling?
 a) Identifying potential drug targets
 b) Designing drugs based on receptor structures
 c) Predicting the biological activity of molecules based on their chemical structure
 d) Analyzing protein–protein interactions

Q10 Which one is best defining the advantage of QSAR modeling in drug discovery?
 a) It does not require experimental validation
 b) It is faster and less expensive than traditional methods
 c) It is more accurate than other drug design approaches
 d) It can predict drug efficacy in clinical trials

**Answers**
Q1  b) Identifying potential binding sites on target proteins
Q2  b) Screening large chemical libraries to identify potential drug candidates
Q3  a) Molecular dynamics
Q4  b) The three-dimensional arrangement of functional groups essential for biological activity
Q5  c) Schrödinger Suite
Q6  d) Pharmacophore modeling
Q7  a) To confirm its effectiveness in predicting drug activity
Q8  c) Chemical properties of molecules
Q9  c) Predicting the biological activity of molecules based on their chemical structure
Q10  b) It is faster and less expensive than traditional methods

Vikas Kushwaha, Anu Prabha, Varruchi Sharma*, Ashwanti Devi,
Seema Ramniwas, Anupam Sharma, Anil K. Sharma*, Imran Sheikh,
Anil Panwar, and Damanjeet Kaur

# Chapter 7
# Immunoinformatics: computational keys to immune system secrets

**Abstract:** Immuno-informatics is an interdisciplinary field that leverages computational approaches to study and understand immune system responses of the host against the pathogens. It a new technology that deals with combinations of immunology with bioinformatics, molecular modelling, and systems biology to analyze large datasets, predict immune responses, and develop new strategies for vaccine design, diagnostics, and immunotherapies. Now, availability of genomic, proteomic, and immunological data, immuno-informatics has become an essential tool in modern immunology research, accelerating the development of new vaccines and therapeutics. The main objectives of immuno-informatics are the prediction of epitopes, which are specific regions on antigens that can be recognized by the immune system, particularly by T-cells and B-cells. Epitope prediction allows for the identification of potential targets for vaccine development, avoiding the need for expensive and time-consuming experimental approaches. Computational tools like machine learning models and molecular docking simulations are widely employed to identify and predict these immunogenic regions from pathogens, cancer cells, or allergens. Additionally, immuno-informatics help in understanding the complex interactions between pathogens and the host im-

*Corresponding author: Varruchi Sharma, Department of Biotechnology and Bioinformatics, Sri Guru Gobind Singh College, Sector 26, Chandigarh, India, e-mail: sharma.varruchi@gmail.com
*Corresponding author: Anil K. Sharma, Department of Biotechnology, Amity School of Biological Sciences, Amity University Punjab, Mohali 140306, Punjab, India, e-mail: anibiotech18@gmail.com
Vikas Kushwaha, Damanjeet Kaur, Department of Biotechnology and Microbial Biotechnology, Sri Guru Gobind Singh College, Sector 26, Chandigarh, India
Anu Prabha, Department of Botany, Goswami Ganesh Dutta Sanatan Dharma College, Sector 32, Chandigarh, India
Ashwanti Devi, Department of Biosciences and Technology, MMEC, Maharishi Markandeshwar (Deemed to be University), Mullana, Ambala, 133207, Haryana, India
Seema Ramniwas, University Centre for Research and Development, University Institute of Biotechnology, Chandigarh University, Gharuan, Mohali, India
Anupam Sharma, Department of Physics, Guru Kashi University, Talwandi Sabo, Bathinda, Punjab, India
Imran Sheikh, Department of Biotechnology, Eternal University, Baru Sahib, Sirmour, Himachal Pradesh, India
Anil Panwar, Department of Bioinformatics and Computational Biology, College of Biotechnology, CCS Haryana Agricultural University, Hisar, Haryana, India

https://doi.org/10.1515/9783111568584-007

mune system. It helps in mapping out the genetic variations in pathogens, immune evasion strategies, and host immune responses, which can be further used to develop personalized immunotherapies and improve the efficacy of vaccines. Overall, immuno-informatics holds significant promise in advancing our understanding of immune responses, driving innovation in disease prevention and treatment through computational tools that enhance the accuracy and efficiency of immunological research.

**Keywords:** Immuno-informatics, B and T cell epitopes, Allergenicity, Antigenicity, IEDB data base, Linkers

# 7.1 Introduction

The human immune system is extremely complicated and functions in coordination with a variety of molecules within cells to protect the organism against various diseases. Based on the conventional dogma of immunology, organisms have both innate and adaptive immune systems that collectively generate immune responses to fight against infections. The innate immune system serves as the first line of defense, swiftly acting as crucial fortifications to protect the body from potential invaders and laying the foundation for the development of adaptive immunity [1]. This broad-spectrum defense mechanism comprises various elements: physical barriers like tight junctions in the skin, epithelial and mucous membranes, and mucus; anatomical barriers; enzymes within epithelial and phagocytic cells such as lysozyme; phagocytic cells like neutrophils, monocytes, and macrophages; inflammation-related serum proteins like complement, C-reactive protein, mannose-binding lectin, and ficolins; antimicrobial peptides present on surfaces and within phagocyte granules like defensins and cathelicidin; Toll-like receptors on cells that detect microorganisms and trigger a defensive reaction; and cytokines and inflammatory mediators released by cells such as macrophages, mast cells, and natural killer cells [2].

Adaptive immune responses, a cornerstone of the body's defense, typically manifest 5–6 days following initial pathogen exposure. These responses hinge on intricate signaling molecules like cytokines and chemokines, orchestrating a sophisticated interplay between immune cells, ultimately culminating in targeted and potent defense mechanisms against attacking pathogens.

The adaptive immune system encompasses cellular responses, mediated by T cells, and humoral responses, mediated by B cells. The epitope of an antigen, recognized by B-cell receptor (BCR) or T-cell receptor (TCR), may be linear or discontinuous for B cells and short linear peptides for T cells. T cells, identified by CD8 or CD4 glycoproteins, consist of CD4 T cells (T helper cells) recognizing peptides displayed by major histocompatibility complex (MHC) class II (MHC-II) molecules and CD8 T cells [3].

Sequencing of human, other organisms, and various human pathogen genomes has provided an extensive amount of data that is crucial for studying human immune diseases and the quest for improved methods for disease prevention or treatment. The immune system is incredibly complex due to its hierarchical and combinatorial nature. This complexity becomes apparent when observing the substantial volume of immunology data that has been amassed. This data, obtained through genomic sequencing, clinical practices, and epidemiological studies, serves as a valuable reservoir for researchers seeking information about immune function mechanisms, immunological interactions, and disease pathogenesis [4]. Immunological researchers store, manage, and analyze immunological data derived from genomic sequencing, scientific practice, and epidemiology. The burgeoning wealth of immunological information finds its management in the emerging discipline of immunoinformatics or computational immunology.

In the 1970s and 1980s, theoretical immunology predominantly utilized mathematical modeling of immune processes. However, this field has grown significantly over the years to include many more areas of immune system informatics, such as molecular and systematic modeling, structural immunoinformatics, and molecular simulations. Analyzing this wealth of immunological data necessitates the use of sophisticated immunoinformatics tools to ensure the precise depiction of immune processes, the discovery of novel insights, and the advancement of disease research applications and treatments. At the moment, progress in immunoinformatics includes making data repositories better organized and structured, creating new computer programs for analyzing immunological data, and creating predictive models to show how different immune systems interact at the molecular, single-cell, and systemic levels.

In the modern realm of immunoinformatics, ongoing efforts are made to enhance the effectiveness of data sources, develop novel instruments for immunological data analysis, and form prediction models that mimic intricate immune relationships. These developments contribute to our understanding of the immune system and have the potential to improve approaches to treating diseases.

## 7.2 Data sources

The primary sources of data in this context include information derived from meticulous laboratory experiments, insights gleaned from the vast realm of scientific literature, a wealth of molecular and immunological databases, dynamic repositories hosted on web servers, comprehensive patient histories that provide valuable contextual information, and meticulous records maintained in clinical settings.

### 7.2.1 Laboratory immunological experiments

Immunological experiments and advanced molecular biology techniques play a pivotal role in generating vast experimental data that aid researchers in comprehending the structure and function of immune genes and their respective products. These experiments employ a diverse array of immunological techniques like affinity chromatography, flow cytometry, radioimmunoassay, and enzyme-linked immunosorbent assay, aimed at understanding the intricate mechanisms of the immune system and its responses to various infections and diseases. The identification and mapping of B- and T-cell epitopes are critical steps in comprehending the complex architecture of immunological interactions. This technique not only helps to understand the underlying processes of immune responses, but it also paves the way for the creation of diagnostic tools and strategies that may successfully identify and treat a wide range of diseases. The rigorous identification and mapping of these epitopes contribute to a comprehensive understanding of immunological dynamics, boosting our ability to diagnose and manage a wide range of medical illnesses with greater precision and efficiency.

In order to predict potential immunodominant epitopes, four antigenic proteins – histone H1, sterol 24-C-methyltransferase, *Leishmania*-specific hypothetical protein, and *Leishmania*-specific antigenic protein – were selected. This multiepitope peptide vaccine stimulates immune responses and protects BALB/c mice against *Leishmania infantum* [5].

The four structural proteins of SARS-CoV-2 contain immunogenic epitopes that were identified and used to develop a multiepitope peptide vaccine that showed high production in *E. coli* and that, when simulated, produced a high level of B-cell- and T-cell-mediated immunity. The findings indicated that administering three shots of the vaccine considerably decreased the growth of the antigen in the system [6]. A multiepitope peptide was synthesized using 8 plasma membrane proteins of *Schistosoma mansoni* that had 19 common MHC-I- and MHC-II-binding epitopes. The purpose of this peptide is to stimulate immunological responses based on B cells and IFN-γ, all the while exhibiting stability and a nonallergic profile. The aim of this multiepitope vaccine is to trigger both humoral and cellular immune responses, positioning it as a formidable candidate for a schistosomiasis vaccine [7].

## 7.2.2 Scientific literature

Scientific literature plays a crucial role as data in immunoinformatics, which is not included in molecular databases. It provides information about various antigens, such as their structures, properties, and interactions with the immune system, as well as data on epitope mapping studies, identifying specific regions of antigens recognized by the immune system. This information is crucial for the prediction of antigenic determinants, designing vaccines, and understanding immune responses [8]. Scientific literature describes immunological pathways, signaling cascades, and molecular interactions, forming the foundation for developing algorithms in immunoinformatics. Researchers curate data to build valuable datasets for tool development and validation. This information is utilized to design and optimize algorithms for tasks such as epitope prediction, protein–protein interaction analysis, and immune system simulation

[9–11]. PubMed holds over 15 million life science abstracts, adding 600,000 records annually, with many linked to full-text articles. The vast information is evident in topic-related additions; for example, in 2004, about 4,200, 6,700, and 17,700 abstracts on MHC or HLA, allergy or allergen, and cytokine, respectively [12]. Universal access to abstracts underscores the need for text-mining tools to extract information. With the growing volume of articles, the quality of immunology research hinges on researchers accessing critical information from data sources.

## 7.2.3 Immunological databases

Immunological databases serve as centralized repositories of information about immunological data that researchers and computational biologists can leverage to gain insights into various aspects of the immune system. Immunological databases are increasing day by day. A total of 45 immunological databases were described in the health science library system in December 2023 (https://www.hsls.pitt.edu/obrc/index.php?page=immunology) (Table 7.1). Recently, the 29th annual Nucleic Acids Research Database Issue contains 185 papers, including 87 papers reporting on new databases, of which 4 are new databases related to immunoinformatics [13]. Immune sequence databases are essential for investigating autoimmune disorders, infectious diseases, cancer, as well as for research in immunotherapy and immunoprophylaxis. IMGT, known as the international ImMunoGeneTics information system, functions as a central repository for information on immunoglobulins (IG), TCRs, MHC, and related proteins within the immune systems of humans and other vertebrates.

## 7.2.4 B-cell epitope databases

Discovering and defining B-cell epitopes within target antigens stands as a pivotal stage in epitope-centric vaccine design, immunodiagnostic assays, and antibody production. B-cell epitopes are categorized into two classes: linear and conformational. Computational methods for predicting these epitopes are categorized accordingly. Linear epitope prediction methods are built on linear sequences and commonly use amino acid propensities (e.g., hydrophilicity, flexibility, and beta turns) for prediction. Modern machine learning models, on the other hand, leverage sequence-derived features like amino acid composition and cooperativity for accurate predictions [14].

There are numerous B-cell epitope-related databases accessible on the Internet and referenced in various papers (Table 7.1). The Protein Data Bank (PDB; https://www.rcsb.org/) database compiles the compounds derived from X-ray crystallography and nuclear magnetic resonance (NMR) experiments and stores the 3D structure of antigens or the antigen–antibody complex. It offers database search and download services, along with detailed information on PDB data file formats [15]. The Immune Epitope Database (IEDB;

https://www.iedb.org/) stands as the foremost and widely utilized database in the prediction of epitopes. IEDB is a comprehensive repository that provides a database of experimentally characterized B-cell epitopes (both linear and conformational), T-cell epitopes, and MHC binding [16]. Bcipep (https://webs.iiitd.edu.in/raghava/bcipep/info.html) provides extensive details on experimentally validated B-cell epitopes along with tools for pinpointing these epitopes on an antigen sequence [17]. The Conformational Epitope Database (CED; http://www.immunet.cn/ced/log.html) compiles B-cell epitopes sourced from literature and those defined through methods such as X-ray diffraction, NMR, scanning mutagenesis, overlapping peptides, and phage display [18]. Epitome (http://www.rostlab.org/services/epitome/) contains information about antigen–antibody complex structures as well as the antibodies that interact with them [19]. AntiJen serves as a comprehensive repository, meticulously curated to amalgamate kinetic, thermodynamic, functional, and cellular information, offering a holistic perspective within the realms of immunology and vaccinology. It provides researchers with a rich resource to explore, analyze, and harness data crucial for advancing understanding and innovation in these fields. The database currently contains experimental data reported in PubMed, quantitative binding data for peptides binding to MHC ligands, TCR–MHC complexes, T-cell epitopes, TAP, B-cell epitope molecules, and immunological protein–protein interactions [20].

## 7.2.5 T-cell epitope databases

T-cell activation necessitates MHC–peptide binding, during which the ligand interacts with a specific TCR. Many databases exist for finding and mapping potential epitopes, which leads to the development of effective vaccines. The IEDB (https://www.iedb.org/) is a resource that is easily accessible to everyone. It serves as a platform for experimental data that characterizes antibody and T-cell epitopes studied in various species, including humans, nonhuman primates, and other animals [21]. A beta version of the IEDB and Analysis Resource Database (http://tools.iedb.org/) is a computational tool focused on the prediction and analysis of B- and T-cell epitopes. The Syfpeithi database (http://www.syfpeithi.de) offers comprehensive information regarding anchor motifs and binding specificity of MHC-I and MHC-II molecules. The score is determined by applying the following rules: computed score values distinguish between anchor, auxiliary anchor, or specified residues [22]. The IMGT is an esteemed integrated information system dedicated to IG, TCR, and MHC data across human and vertebrate species. IMGT encompasses three sequence databases (IMGT/LIGM-DB, IMGT/MHC-DB, and IMGT/PRIMER-DB), a genome database (IMGT-GENE-DB), a 3D structure database (IMGT/3Dstructure-DB), along with various interactive tools for sequence analysis. Accessible through different interfaces like IMGT/GeneSearch and IMGT/PhyloGene, IMGT offers a wealth of resources for researchers in immunology and beyond [23].

# 7.3 Immunological tools

The term "immunogenicity" refers to the capacity of antigens to bind specifically to antibodies and initiate immune responses within the body. In the context of immunoinformatics, the prediction and design of epitopes aim to optimize the immunogenicity of these molecules. This involves the strategic selection and construction of epitopes to elicit robust and effective immune responses against a given antigen. The goal is to enhance the antigen's ability to stimulate the immune system, ultimately contributing to the development of vaccines or therapeutic interventions with heightened efficacy. The careful consideration and prediction of epitopes play a pivotal role in tailoring immune responses for improved protection or treatment outcomes.

Various in silico methodologies are currently under development and in use for the identification of epitopes. These tools encompass prediction techniques for B- and T-cell epitopes, allergy prediction, and in silico vaccination.

## 7.3.1 B-cell epitope prediction

The identification and mapping of B-cell antigen epitopes are critical steps in the development of effective immunogens that aid in the production of specific antibodies. B-cell epitopes possess the unique ability to bind to IGs and BCRs, forming the basis for the elicitation of specific immune responses. Traditional physical methods, such as crystallography-based techniques, mass spectrometry-based methods, NMR, and surface plasmon resonance, are effective but often time-consuming and expensive for the determination of B-cell epitopes [24].

In contrast, computational methods have emerged as a reliable and efficient alternative for the identification of B-cell epitopes. These computational tools leverage algorithms and predictive models to analyze the antigen's sequence and predict regions likely to elicit an immune response. By offering a quicker and more cost-effective approach, computational methods complement traditional techniques, contributing to the advancement of epitope mapping and accelerating the immunogen design process for antibody production.

B-cell epitopes are classified into two types: linear, also known as continuous, and conformational, also referred to as discontinuous. Linear epitopes are composed of short peptides arranged in sequential amino acid order within the antigen. In contrast, discontinuous epitopes consist of antigen residues that are physically distant in the primary sequence but are brought into close proximity through the folding of the polypeptide chain [4].

B-cell epitope prediction mostly uses a number of different methods, such as sequence-based analysis, machine learning algorithms, and amino acid propensity scale-based approaches. Sequence-based methods focus on surface accessibility for antibody binding, predicting continuous epitopes [25]. Amino acid propensity scale-based methods establish a correlation between the positions of continuous epitopes and attributes such as accessibility, hydrophilicity, flexibility, exposed surface, polarity turns, and antigenic propensity. Amino acid propensity scales are essential for predicting linear B-cell epitopes [26]. Currently, both amino acid scales and machine learning approaches have demonstrated the highest efficacy in predicting continuous epitopes. Improving the quality of current B-cell epitope datasets would lead to enhanced and more accurate outcomes. Numerous prediction tools for B-cell epitopes are now accessible for various antigens. These include ABCpred [27], BciPep [17], Bepipred [28], IMGT® [29], Bcepred [30], BEPITOPE [31], DiscoTope [32], COBEpro [33], CEP [34], IEDB [35], AntiJen [36], and Pepitope [37] (Table 7.1).

## 7.3.2 T-cell epitope prediction

A T-cell epitope is a short peptide sequence from an antigen that is presented on the surface of an antigen-presenting cell (APC) and recognized by a TCR. These epitopes are bound to MHC molecules, which are displayed on the cell surface of APCs. MHC-I and MHC-II molecules are identified by CD8 and CD4 T-cell epitope subsets, respectively [38].

In order to create an epitope-based vaccine, it is necessary to first identify T-cell epitopes and then bind them with the correct HLA molecules. To recognize cytotoxic T cells, antigenic peptides must be molecularly bound to MHC molecules. T-cell prediction tools are computational resources that utilize algorithms, machine learning, and bioinformatics techniques to analyze protein sequences and predict potential T-cell epitopes [39]. Various approaches, such as hidden Markov model, artificial neural networks, support vector machine (SVM), and quantitative matrices, can be used to forecast peptides that have the ability to bind MHC molecules [40]. Various T-cell epitope prediction tools are now available, including IEDB [35], SYFPEITHI [22], MHCPred [41], NetCTL [42], NetMHC [43], EpiToolKit [44], CTLpred [45], and PREDEPP [46] (Table 7.2).

## 7.3.3 Linkers

In the context of epitope vaccine design, linkers are sequences of amino acids used to join different epitopes or functional domains that elicit specific immune responses. Linkers can be designed to optimize spacing and orientation between epitopes, allow flexibility, proper folding, and effective presentation of individual components of epitopes, enhancing the immune response. Linkers are categorized as flexible, rigid, cleavable, and enriched in small, polar amino acids like glycine and serine, providing solubility and flexibility. Notably, flexible linkers can also act as passive spacers to maintain a distance between functional domains [47, 48].

The following four linkers, viz. EAAAK, GPGPG, AAY, and KK, were basically used for joining the epitopes [49].

Linkers present a myriad of benefits in crafting fusion proteins, such as enhancing biological activity, fine-tuning pharmacokinetics, and boosting expression yields. Their strategic incorporation holds the promise of unlocking new avenues for creating more potent and versatile therapeutic agents with enhanced efficacy and functionality.

The LINKER database enables users to explore linker peptides using different query types, delivering matching sequences. Furthermore, it offers an option to generate linker sequences derived from experimentally resolved structures, favoring extended conformations suitable for fusion proteins. Another online tool, the Linker Database at Vrije University of Amsterdam's Center for Integrative Bioinformatics, presents a search engine that filters results based on diverse conformations. The platform furnishes linker sequences aligning with search criteria, alongside key information such as PDB code, source protein description, linker length, solvent accessibility, linker position within the source protein, and linker secondary structure, facilitating comprehensive exploration and selection of suitable linkers for fusion protein design [50].

## 7.3.4 Allergenicity and antigenicity prediction tool

Allergen is a protein or glycoprotein recognized by IgE and produces a vigorous immune response that causes an allergic reaction in all age groups. Various online allergen databases and allergy prediction tools are being used to find the allergenicity. AllergenPro (http://nabic.rda.go.kr/allergen/) is an integrated web-based database for allergenicity analysis [51]. AllergenOnline (http://www.allergenonline.org/) offers a peer-reviewed allergen list and a searchable database for identifying proteins with potential allergenic cross-reactivity. Antigenicity is a property of antigens that allows them to bind to components of the adaptive immune system [52]. IEDB (https://www.iedb.org/) and SVMTriP (http://sysbio.unl.edu/SVMTriP/) are a few important databases that help in the identification of antigenicity of antigens. The list of the most reliable antigenicity and allergenicity prediction tools is presented in Table 7.3.

## 7.3.5 Design optimization

Design optimization includes the evaluation of posttranslational modifications (PTMs) and codon optimization of the designed antigen. PTMs, such as glycosylation, phosphorylation, and palmitoylation, occur on the C- or N-termini of amino acid side chains, impacting protein structure, dynamics, and biological activity. Various computational tools are available to predict PTM sites within protein sequences, and a comprehensive list of these tools is provided in Table 7.4.

Codon optimization is the process of strategically modifying the nucleotide sequence of a gene without altering the encoded amino acid sequence. During codon optimization, synonymous codons, which code for the same amino acid but have different nucleotide sequences, may be selected based on the host organism's codon usage bias. This optimization aims to improve translation efficiency, mRNA stability, and overall protein expression levels in the host system. The process is particularly relevant in recombinant DNA technology and synthetic biology, where genes from one organism are introduced into another for expression. By aligning the codon usage patterns of the foreign gene with those of the host organism, codon optimization can help overcome potential barriers related to translation kinetics and enhance the overall success of heterologous gene expression. Various codon optimization tools have been developed to enhance the expression of recombinant proteins. Table 7.5 presents a list of the most reliable codon optimization tools.

## 7.4 Conclusion

Immunoinformatics emerges as an exciting frontier, employing cutting-edge computational methods to decipher the complexities of immune responses and their interplay with pathogens or treatments. This rapidly advancing field holds immense promise for revolutionizing our understanding of immunity, paving the way for groundbreaking therapeutic interventions and disease management strategies. The integration of bioinformatics, computational biology, and immunology has led to significant advancements in the design and optimization of vaccines, immunotherapies, and diagnostic tools. Immunoinformatics plays a crucial role in epitope prediction, facilitating the identification of potential antigenic determinants for vaccine development with enhanced efficacy. Additionally, the field contributes to our understanding of immune responses at a systems level, aiding in the identification of immune correlates of protection and potential biomarkers.

The application of immunoinformatics extends beyond vaccine development, encompassing the prediction of T-cell and B-cell epitopes, MHC binding, and molecular docking studies. These computational approaches expedite the screening and selection of candidate antigens and improve our understanding of the immune system's recognition and response mechanisms. Furthermore, as technologies advance, immunoinformatics is poised to contribute significantly to personalized medicine by tailoring immunotherapies based on individual genetic and immunological profiles.

In summary, immunoinformatics continues to be a dynamic and indispensable field in the quest for innovative and effective immunological interventions. The synergy between experimental and computational approaches is vital for advancing our understanding of immune responses, ultimately leading to the development of more targeted and efficacious immunotherapies and vaccines.

**Table 7.1:** List of B-cell epitope prediction tools.

| Tools | Web addresses |
|---|---|
| ABCpred | http://www.imtech.res.in/raghava/abcpred |
| BCIPEP | http://www.imtech.res.in/raghava/bcipep |
| Bepipred | http://www.cbs.dtu.dk/services/BepiPred |
| IMGT VR | http://www.imgt.org |
| Bcepred | http://www.imtech.res.in/raghava/bcepred/ |
| BEPITOPE | jlpellequer@cea.fr |
| COBEpro | http://scratch.proteomics.ics.uci.edu |
| CEP | http://bioinfo.ernet.in/cep.htm |
| MIMOX | http://www.immunet.cn/mimox/ |
| Pepitope | http://www.pepitope.tau.ac.il/ |
| 3DEX | http://www.schreiber-abc.com/3dex/ |
| IEDB | http://www.immuneepitope.org |
| AntiJen | http://www.jenner.ac.uk/antijen/ |
| CED | http://immunet.cn/ced |

**Table 7.2:** List of T-cell epitope prediction tools.

| Tools | Web addresses |
|---|---|
| Allele frequencies | http://www.allelefrequencies.net |
| MHCPred | http://www.ddg-pharmfac.net/mhcpred/MHCPred/ |
| MMBPred | http://www.imtech.res.in/raghava/mmbpred/ |
| NetCTL | http://www.cbs.dtu.dk/services/NetCTL |
| NetMHC | http://www.cbs.dtu.dk/services/NetMHC |
| NetChop | http://www.cbs.dtu.dk/services/NetChop |
| TAPPred | http://www.imtech.res.in/raghava/tappred/ |
| Pcleavage | http://www.imtech.res.in/raghava/pcleavage/ |
| Propred | http://www.imtech.res.in/raghava/propred1/ |
| ElliPro | http://www.tools.immuneepitope.org/tools/ElliPro |
| EpiToolKit | http://www.epitoolkit.org |
| EpiVax | http://www.epivax.com |
| MAPPP | www.mpiib-berlin.mpg.de/MAPPP/cleavage.html |
| SYFPEITHI | http://www.syfpeithi.de |
| IMGT VR | http://www.imgt.org |
| IMGT/HLA | http://www.ebi.ac.uk/imgt/hla/allele.html |
| IEDB | http://www.immuneepitope.org |
| EpiJen v 1.0 | http://www.ddg-harmfac.net/epijen/EpiJen/EpiJen.htm |
| BIMAS | http://www.thr.cit.nih.gov/molbio/hla_bind |
| CTLpred | http://www.imtech.res.in/raghava/ctlpred |
| JenPep | http://www.jenner.ac.uk/JenPep |
| PREDEPP | http://margalit.huji.ac.il/Teppred/mhc-bind/index.html |
| PDB | https://www.rcsb.org/pdb/ MHC/peptide/TCR combinations |
| TEPITOPE | https://www.vaccinome.com |

**Table 7.3:** List of antigenicity and allergenicity prediction tools.

| Tools | Web addresses |
|---|---|
| AllergenPro | http://nabic.rda.go.kr/allergen/ |
| Allergome | www.allergome.org |
| AlgPred | http://www.imtech.res.in/raghava/algpred/ |
| Allermatch | http://www.allermatch.org |
| AllerTOP 2.0 | http://www.ddg-pharmfac.net/AllerTOP |
| ANTIGENpro | http://scratch.proteomics.ics.uci.edu/ |
| APPEL | http://jing.cz3.nus.edu.sg/cgi-bin/APPEL |
| Database of IUIS | http://www.allergen.org |
| FARRP | https://farrp.unl.edu/ |
| SDAP | http://fermi.utmb.edu/SDAP/ |
| VaxiJen | www.ddg-pharmfac.net/vaxijen/ |

**Table 7.4:** List of PTM prediction tools.

| Type of PTM | Tools | Web addresses |
|---|---|---|
| Phosphorylation | NetPhos | http://www.cbs.dtu.dk/services/NetPhos/ |
| | PhosphoELM | http://phospho.elm.eu.org |
| | Scansite | http://scansite.mit.edu |
| | Phosphositeplus | http://www.phosphosite.org/ |
| | GPS | http://gps.biocuckoo.org/online.php |
| | KinasePhos 2.0 | http://kinasephos2.mbc.nctu.edu.tw |
| | PPRED | http://biomecis.uta.edu/ashis/res/ppred/ |
| Glycosylation | NetNGlyc1.0 | www.cbs.dtu.dk/services/NetNGlyc/ |
| | NetOGlyc 4.0 | www.cbs.dtu.dk/services/NetOGlyc/ |
| | GlycoDomain Viewer | http://glycodomain.glycomics.ku.dk/ |
| | O-GlcNAcPRED | http://121.42.167.206/OGlcPred/ |
| | GlycoMine | http://www.structbioinfor.org/Lab/GlycoMine/ |
| | GlySeq | http://www.glycosciences.de/tools/glyseq/instructions.php |
| Palmitoylation | CSS-Palm 4.0 | http://csspalm.biocuckoo.org/ |
| | NBA-Palmj | http://www.bioinfo.tsinghua.edu.cn/NBA-Palm |
| | WAP-Palm | http://bioinfo.ncu.edu.cn/WAP-Palm.aspx |
| Methylation | iSNO-PseAAC | http://app.aporc.org/iSNO-PseAAC/ |
| | MeMo | http://www.bioinfo.tsinghua.edu.cn/ tigerchen/memo.html |
| | BPB-PPMS | http://www.bioinfo.bio.cuhk.edu.hk/bpbppms/ |
| | MASA | http://masa.mbc.nctu.edu.tw/ |
| | PMes | http://bioinfo.ncu.edu.cn/inquiries_PMeS.aspx |
| | MethK | http://csb.cse.yzu.edu.tw/MethK |
| | iMethyl-PseAAC | http://www.jci-bioinfo.cn/iMethyl-PseAAC |
| N-acetylation | PSSMe | http://bioinfo.ncu.edu.cn/PSSMe.aspx |
| | NetAcet | http://www.cbs.dtu.dk/services/NetAcet |
| | PAIL | http://bdmpail.biocuckoo.org/prediction.php |
| | N-Ace | http://n-ace.mbc.nctu.edu.tw |

**Table 7.5:** List of codon optimization tools.

| Tools | Web addresses |
| --- | --- |
| Codon usage database | http://www.kazusa.or.jp/codon/ |
| COOL | http://bioinfo.bti.a-star.edu.sg/COOL/ |
| GenScript | https://www.genscript.com/codon-opt.html |
| Integrated DNA Technologies, Inc | https://eu.idtdna.com/CodonOpt# |
| Jcat | www.jcat.de/ |
| Optimizer | http://genomes.urv.es/OPTIMIZER/ |

# References

[1]   The innate and adaptive immune systems. In: InformedHealth.org [Internet] [Internet]. Institute for Quality and Efficiency in Health Care (IQWiG); 2020 [cited 2024 Feb 3]. Available from: https://www.ncbi.nlm.nih.gov/books/NBK279396/

[2]   Aristizábal B., González Á. Innate Immune System. In Autoimmunity: From Bench to Bedside [Internet] [Internet], El Rosario University Press; 2013, [cited 2024 Feb 3]. Available from. https://www.ncbi.nlm.nih.gov/books/NBK459455/.

[3]   Alberts B., Johnson A., Lewis J., Raff M., Roberts K., Walter P. The Adaptive Immune System. In Molecular Biology of the Cell, 4th Edn, [Internet], Garland Science; 2002, [cited 2024 Feb 3]. Available from. https://www.ncbi.nlm.nih.gov/books/NBK21070/.

[4]   Tong J.C., Ren E.C. Immunoinformatics: Current trends and future directions. *Drug Discovery Today: Technologies*, 2009 Jul. **14**(13–14): 684–689.

[5]   Vakili B., Nezafat N., Zare B., Erfani N., Akbari M., Ghasemi Y. et al. A new multi-epitope peptide vaccine induces immune responses and protection against Leishmania infantum in BALB/c mice. *Medical Microbiology and Immunology (Berl)*, 2020 Feb. **209**(1): 69–79.

[6]   Singh A., Thakur M., Sharma L.K., Chandra K. Designing a multi-epitope peptide based vaccine against SARS-CoV-2. *Scientific Reports*, 2020 Oct 1. **10**(1): 16219.

[7]   Sanches R.C.O., Tiwari S., Ferreira L.C.G., Oliveira F.M., Lopes M.D., Passos M.J.F. et al. Immunoinformatics design of multi-epitope peptide-based vaccine against Schistosoma mansoni using transmembrane proteins as a target. *Frontiers in Immunology*, 2021. **12**: 621706.

[8]   Tomar N., De R.K. Immunoinformatics: A brief review. *Methods in Molecular Biology*, 2014. **1184**: 23–55.

[9]   Razali S.A., Shamsir M.S., Ishak N.F., Low C.F., Azemin W.A. Riding the wave of innovation: Immunoinformatics in fish disease control. *Peer Journal*, 2023 Dec 8. **11**: e16419.

[10]  Bhattacharya M., Sharma A.R., Mallick B., Sharma G., Lee S.S., Chakraborty C. Immunoinformatics approach to understand molecular interaction between multi-epitopic regions of SARS-CoV-2 spike-protein with TLR4/MD-2 complex. *Infection Genetics & Evolution*, 2020 Nov. **85**: 104587.

[11]  Espinosa-Riquer Z.P., Segura-Villalobos D., Ramírez-Moreno I.G., Pérez Rodríguez M.J., Lamas M., Gonzalez-Espinosa C. Signal transduction pathways activated by innate immunity in mast cells: Translating sensing of changes into specific responses. *Cells*, 2020 Nov 4, 9(11): 2411.

[12]  Brusic V., Petrovsky N. Immunoinformatics and its relevance to understanding human immune disease. *Expert Review of Clinical Immunology*, 2005 May. **1**(1): 145–157.

[13]  Rigden D.J., Fernández X.M. The 2023 nucleic acids research database issue and the online molecular biology database collection. *Nucleic Acids Research*, 2023 Jan 6. **51**(D1): D1–8.

[14]  Liu J., Zhang W. Databases for B-cell epitopes. *Methods in Molecular Biology*, 2014. **1184**: 135–148.

[15] Berman H.M., Westbrook J., Feng Z., Gilliland G., Bhat T.N., Weissig H. et al. The protein Data Bank. *Nucleic Acids Research*, 2000 Jan 1. **28**(1): 235–242.

[16] Vita R., Mahajan S., Overton J.A., Dhanda S.K., Martini S., Cantrell J.R. et al. The Immune Epitope Database (IEDB): 2018 update. *Nucleic Acids Research*, 2019 Jan 8. **47**(D1): D339–43.

[17] Saha S., Bhasin M., Raghava GP. Bcipep: A database of B-cell epitopes. *BMC Genomics*, 2005 May 29. **6**(1): 79.

[18] Huang J., Honda W. CED: A conformational epitope database. *BMC Immunology*, 2006 Apr 7. **7**(1): 7.

[19] Schlessinger A., Ofran Y., Yachdav G., Rost B. Epitome: Database of structure-inferred antigenic epitopes. *Nucleic Acids Research*, 2006 Jan 1. **34**(Database issue): D777–80.

[20] Toseland C.P., Clayton D.J., McSparron H., Hemsley S.L., Blythe M.J., Paine K. et al. AntiJen: A quantitative immunology database integrating functional, thermodynamic, kinetic, biophysical, and cellular data. *Immunome Research*, 2005 Oct 6. **1**(1): 4.

[21] Sathiamurthy M., Peters B., Bui H.H., Sidney J., Mokili J., Wilson S.S. et al. An ontology for immune epitopes: Application to the design of a broad scope database of immune reactivities. *Immunome Research*, 2005 Sep 20. **1**: 2.

[22] Rammensee H., Bachmann J., Emmerich N.P., Bachor O.A., Stevanović S. SYFPEITHI: Database for MHC ligands and peptide motifs. *Immunogenetics*, 1999 Nov. **50**(3–4): 213–219.

[23] Lefranc M.P., Giudicelli V., Ginestoux C., Bodmer J., Müller W., Bontrop R. et al. IMGT, the international ImMunoGeneTics database. *Nucleic Acids Research*, 1999 Jan 1. **27**(1): 209–212.

[24] Ahmad T.A., Eweida A.E., Sheweita S.A. B-cell epitope mapping for the design of vaccines and effective diagnostics. *Trials in Vaccinology*, 2016 Jan 1. **5**: 71–83.

[25] Ras-Carmona A., Lehmann A.A., Lehmann P.V., Reche P.A. Prediction of B cell epitopes in proteins using a novel sequence similarity-based method. *Scientific Reports*, 2022 Aug 12. **12**(1): 13739.

[26] Liu T., Shi K., Li W. Deep learning methods improve linear B-cell epitope prediction. *BioData Mining*, 2020 Apr 17. **13**(1): 1.

[27] Sun P., Ju H., Liu Z., Ning Q., Zhang J., Zhao X. et al. Bioinformatics resources and tools for conformational B-cell epitope prediction. *Computational and Mathematical Methods in Medicine*, 2013. 2013: 943636.

[28] Larsen J.E.P., Lund O., Nielsen M. Improved method for predicting linear B-cell epitopes. *Immunome Research*, 2006. **2**: 2.

[29] Lefranc M.P., Giudicelli V., Ginestoux C., Jabado-Michaloud J., Folch G., Bellahcene F. et al. IMGT(R), the international ImMunoGeneTics information system(R). *Nucleic Acids Research*, 2008 Oct. **37** (suppl. 1): D1006–12.

[30] Saha S., GPS R. BcePred: Prediction of Continuous B-Cell Epitopes in Antigenic Sequences Using Physico-chemical Properties. In Nicosia G., Cutello V., Bentley P.J., Timmis J. Eds, Artificial Immune Systems, Berlin, Heidelberg: Springer, 2004, 197–204. Lecture Notes in Computer Science.

[31] Odorico M., Pellequer J.L. BEPITOPE: Predicting the location of continuous epitopes and patterns in proteins. *Journal of Molecular Recognition JMR*, 2003. **16**(1): 20–22.

[32] Haste Andersen P., Nielsen M., Lund O. Prediction of residues in discontinuous B-cell epitopes using protein 3D structures. *Protein Science*, 2006 Nov. **15**(11): 2558–2567.

[33] Sweredoski M.J., Baldi P. COBEpro: A novel system for predicting continuous B-cell epitopes. *Protein Engineering, Design & Selection*, 2009 Mar. **22**(3): 113–120.

[34] CEP: A conformational epitope prediction server | Nucleic Acids Research | Oxford Academic [Internet]. [cited 2024 Feb 3]. Available from: https://academic.oup.com/nar/article/33/suppl_2/W168/2505653

[35] Vita R., Zarebski L., Greenbaum J.A., Emami H., Hoof I., Salimi N. et al. The immune epitope database 2.0. *Nucleic Acids Research*, 2010 Jan. **38**(Database issue): D854–862.

[36]  Toseland C.P., Clayton D.J., McSparron H., Hemsley S.L., Blythe M.J., Paine K. et al. AntiJen: A quantitative immunology database integrating functional, thermodynamic, kinetic, biophysical, and cellular data. *Immunome Research*, 2005 Oct 6. **1**(1): 4.

[37]  Mayrose I., Penn O., Erez E., Rubinstein N.D., Shlomi T., Freund N.T. et al. Pepitope: Epitope mapping from affinity-selected peptides. *Bioinformatics*, 2007 Dec 1. **23**(23): 3244–3246.

[38]  Sanchez-Trincado J.L., Gomez-Perosanz M., Reche P.A. Fundamentals and methods for T- and B-cell epitope prediction. *Journal of Immunology Research*, 2017. 2017: 2680160.

[39]  Patronov A., Doytchinova I. T-cell epitope vaccine design by immunoinformatics. *Open Biology*, 2013 Jan 8. **3**(1): 120139.

[40]  Schaap-Johansen A.L., Vujović M., Borch A., Hadrup S.R., Marcatili P. T cell epitope prediction and its application to immunotherapy. *Frontiers in Immunology*, [Internet]. 2021 [cited 2024 Feb 3];12. Available from. https://www.frontiersin.org/articles/10.3389/fimmu.2021.712488.

[41]  Guan P., Doytchinova I.A., Zygouri C., Flower D.R. MHCPred: A server for quantitative prediction of peptide–MHC binding. *Nucleic Acids Research*, 2003 Jul 1. **31**(13): 3621–3624.

[42]  Heiny A.T., Miotto O., Srinivasan K.N., Khan A.M., Zhang G.L., Brusic V. et al. Evolutionarily conserved protein sequences of influenza a viruses, avian and human, as vaccine targets. *PLOS One*, 2007 Nov21. **2**(11): e1190.

[43]  NetMHC-3.0: Accurate web accessible predictions of human, mouse and monkey MHC class I affinities for peptides of length 8–11–PubMed [Internet]. [cited 2024 Feb 3]. Available from: https://pubmed.ncbi.nlm.nih.gov/18463140/

[44]  EpiToolKit – A web server for computational immunomics | Nucleic Acids Research | Oxford Academic [Internet]. [cited 2024 Feb 3]. Available from: https://academic.oup.com/nar/article/36/suppl_2/W519/2506015

[45]  Bhasin M., Raghava G.P.S. Prediction of CTL epitopes using QM, SVM and ANN techniques. *Vaccine*, 2004 Aug13. **22**(23–24): 3195–3204.

[46]  Lundegaard C., Lund O., Buus S., Nielsen M. Major histocompatibility complex class I binding predictions as a tool in epitope discovery. *Immunology*, 2010 Jul. **130**(3): 309–318.

[47]  Chen X., Zaro J.L., Shen W.C. Fusion protein linkers: Property, design and functionality. *Advanced Drug Delivery Reviews*, 2013 Oct. **65**(10): 1357–1369.

[48]  Savsani K., Jabbour G., Dakshanamurthy S. A new epitope selection method: Application to design a multi-valent epitope vaccine targeting HRAS oncogene in squamous cell Carcinoma. *Vaccines (Basel)*, 2021 Dec 31. **10**(1): 63.

[49]  Design of a multi-epitope-based vaccine targeting M-protein of SARS-CoV2: An immunoinformatics approach – PubMed [Internet]. [cited 2024 Feb 3]. Available from: https://pubmed.ncbi.nlm.nih.gov/33252008/

[50]  George R.A., Heringa J. An analysis of protein domain linkers: Their classification and role in protein folding. *Protein Engineering*, 2002 Nov. **15**(11): 871–879.

[51]  Neeharika D., Sunkar S. Computational approach for the identification of putative allergens from Cucurbitaceae family members. *Journal of Food Science and Technology*, 2021 Jan. **58**(1): 267–280.

[52]  (99+) Kuby Immunology 7th c2013 txtbk | Paulinha Pereira – Academia.edu [Internet]. [cited 2024 Feb 3]. Available from: https://www.academia.edu/40436029/Kuby_Immunology_7th_c2013_txtbk

**Multiple choice questions**

Q1   What is the primary goal of immunoinformatics?
   a)   Analyzing gene expression levels in immune cells
   b)   Designing vaccines and immunotherapies
   c)   Predicting protein structures in the immune system
   d)   Studying immune cell signaling pathways

Q2   Which is the most common application of immunoinformatics?
   a)   Identifying potential drug targets in cancer cells
   b)   Analyzing protein–protein interactions in the brain
   c)   Predicting the secondary structure of RNA molecules
   d)   Designing peptide vaccines for infectious diseases

Q3   Which computational technique is used for epitope prediction?
   a)   Molecular docking
   b)   Hidden Markov models
   c)   Machine learning algorithms
   d)   Gel electrophoresis

Q4   Why epitopes are necessary in immunoinformatics? Because
   a)   They are antigens that stimulate immune responses
   b)   They are antibodies produced by the immune system
   c)   They are receptors on immune cells
   d)   They are cytokines secreted by immune cells

Q5   T-cell epitopes predicted by?
   a)   By analyzing protein–protein interactions
   b)   By predicting peptide binding to MHC molecules
   c)   By analyzing gene expression levels in immune cells
   d)   By predicting RNA secondary structures

Q6   Which database is used in immunoinformatics for antigen retrieval?
   a)   NCBI Gene
   b)   Swiss-Prot
   c)   Immune Epitope Database (IEDB)
   d)   Protein Data Bank (PDB)

Q7   What is the purpose of reverse vaccinology?
   a)   To design vaccines based on protein structures
   b)   To analyze immune responses in vaccinated individuals
   c)   To identify potential drug targets in pathogens
   d)   To study the immune system's response to infection

Q8   Which bioinformatics tool is used for protein structure prediction in immunoinformatics?
   a)   BLAST
   b)   Clustal Omega
   c)   I-TASSER
   d)   Primer3

Q9 How does immunoinformatics contribute to personalized medicine?
  a) By predicting drug responses based on genetic information
  b) By analyzing protein–protein interactions in cancer cells
  c) By designing customized vaccines based on individual immune profiles
  d) By studying the role of the immune system in autoimmune diseases

Q10 Which is the codon optimization tool?
  a) GenScript
  b) Pred
  c) NetMHC
  d) IEDB

**Answers**

Q1  b)  Designing vaccines and immunotherapies
Q2  d)  Designing peptide vaccines for infectious diseases
Q3  c)  Machine learning algorithms
Q4  a)  They are antigens that stimulate immune responses
Q5  b)  By predicting peptide binding to MHC molecules
Q6  c)  Immune Epitope Database (IEDB)
Q7  a)  To design vaccines based on protein structures
Q8  c)  I-TASSER
Q9  c)  By designing customized vaccines based on individual immune profiles
Q10 a)  GenScript

Vikas Kumar and Nitin Sharma*

# Chapter 8
# Phylogenetic analysis

**Abstract:** Phylogeny refers to the evolutionary origin of organisms. Phylogenetics deals with the investigation of phylogenies, which serve as representations of the evolutionary connections between different species. This tool facilitates the assessment of evolutionary links among different species. Molecular phylogenetic analysis utilizes the sequence of a prevalent gene or protein to evaluate the evolutionary correlation among species. The phylogenetic tree or a branching, tree-like figure is typically used to represent the evolutionary relationship that phylogenetic research reveals. In the past, phylogenetic trees were primarily confined to examining evolutionary biology and related fields such as systematics and taxonomy. This chapter elucidates the fundamental significance of evolutionary links, which may be effectively established through phylogenetic analysis. This analytical approach is pivotal in investigating evolutionary biology, systematics, taxonomy, and contemporary drug development.

**Keywords:** Phylogenetics, phylogenetic analysis, molecular phylogenetic analysis, evolutionary biology, taxonomy, drug development

## 8.1 Introduction

Phylogenetic trees have permeated many fields of biology and beyond since the invention of sequencing and the widespread application of cladistics. The construction of phylogenetic or evolutionary trees has become widely utilized across various disciplines to investigate and illustrate evolutionary divergence. Phylogenetics is the foundation for comparative genomics, a relatively new concept that has emerged in the genomics era. Comparative genomics is an academic discipline investigating the interrelationships between genomes of different species. Comparative genomics facilitates the identification of genomic areas that exhibit similarities and differences across many genomes. The comparison can occur at various levels, depending on what is being compared, such as whole-genome sequences, genome sequences with conserved synteny segments, the number of protein-coding genes, regulatory sequences, or other oriented comparisons. The identification of genes is a fundamental approach within the

*Corresponding author: Nitin Sharma, Department of Biotechnology, Chandigarh Group of Colleges, Landran, Mohali, Punjab, India, e-mail: abhinitu30@gmail.com
Vikas Kumar, Department of Biotechnology, University Institute of Biotechnology, Chandigarh University, Gharuan, Mohali 140413, Punjab, India

https://doi.org/10.1515/9783111568584-008

field of comparative genomics. The area of comparative genomics contributes to the comprehension of evolutionary links among genomes within the context of evolutionary biology.

## 8.2 Phylogenetic tree and its construction

A phylogenetic or evolutionary tree visually illustrates the evolutionary connections between many taxa [1]. The structure consists of interconnected nodes and branches, resulting in a branching configuration referred to as the tree's topology. The nodes within the system symbolize taxonomic entities, which can encompass species, populations, genes, or proteins. An edge branch represents the estimated time between taxonomic units in the context of evolutionary relationships. A single branch can only connect two nodes. The terminal nodes represent the taxonomic operational units (OTUs) or leaves within a phylogenetic tree. The subjects of comparison in this study are referred to as OTUs, encompassing various entities such as species, populations, or gene and protein sequences. Simultaneously, the internal nodes represent inferred hypothetical taxonomic units , the last common ancestor (LCA), wherein subsequent nodes originate. Sister groups are formed by taxa that diverge from a common node, while a taxon that falls outside the clade is often referred to as an outgroup (Figure 8.1).

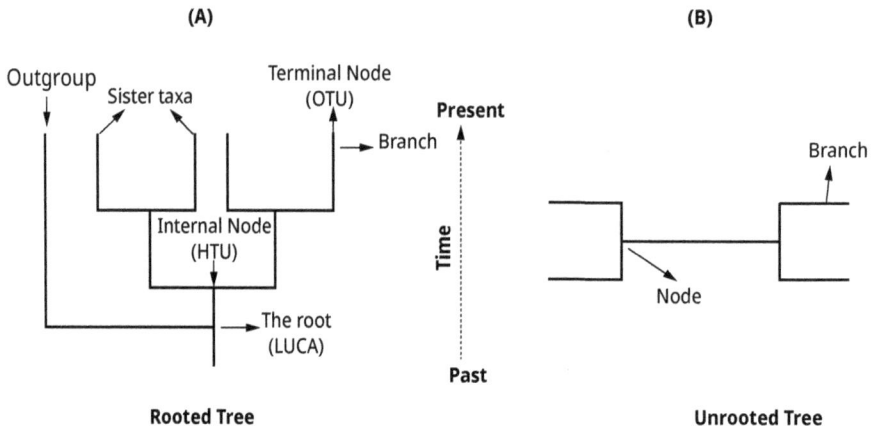

**Figure 8.1:** Different forms of the phylogenetic tree.

Phylogenetic trees can be categorized under two main categories: distance-based and character-based, also called discrete methods. However, choosing a suitable tree-building method for a given dataset is very important. Several methods for reconstructing phylogenetic trees have been described, each with its merits and demerits [2].

A) **Distance-based (distance-matrix) methods:** Distance-based approaches create a distance matrix by measuring the distance between pairs of sequences to build trees. Substitution models are the foundation of distance-based strategies. UPGMA (unweighted pair group method with arithmetic mean) and NJ (neighbor joining) are distance-based approaches. A rooted phylogenetic tree can be constructed with the help of UPGMA, the simplest distance-matrix method, employing successive clustering. Pairwise alignment is first used to compare all sequences and generate a distance matrix. This matrix is used to find the two sequences that are the least far from one another and place them in the same cluster. The distance between this pair and all other sequences is then recalculated to produce a new matrix. The new matrix is used to find and group together sequences most similar to the initial two. This is done repeatedly until all sequences have been added to the cluster. The most popular approach to working with distance matrices is NJ [3]. The process begins with a star tree, a family tree in which the branches leading to various OTUs (the sequences) spread outward from a central node. Next, a pair of sequences is chosen randomly, extracted from the star, and joined to a second internal node through a branch leading back to the star's center. After determining the total length of the branches, the two sequences return to their starting points and select another set of nodes. This approach is repeated to identify the neighborhood combination, yielding the shortest possible phylogenetic tree until all potential pairs have been investigated.

B) **Character-based methods:** In contrast to distance-matrix methods, character-based approaches depend on the intrinsic sequence properties rather than the pairwise distances produced from these qualities. A character can be defined as a distinct position or spatial arrangement within the alignment. There are two extensively established character-based techniques, specifically maximum parsimony (MP) and maximum likelihood (ML) (Figure 8.2). The MP method calculates many trees based on the provided dataset and assigns a cost to each tree. The principle of MP posits that the tree with the fewest evolutionary changes is the most likely. The simplest tree is the one that requires a substantially reduced number of alignment modifications to account for the observed outcomes. Thus, parsimony uses the data rather than attempting to estimate the total number of changes using a model. The score assigned to the tree corresponds to the sum of the character lengths across all positions. The trees are equally cost-effective if more than one can be acquired with the fewest alterations. A genomic locus that exhibits a minimum of two distinct nucleotide types (bases) throughout a minimum of two sequences is commonly referred to as an informative site in the context of MP analysis. The ML method is a statistical approach used to estimate the unknown parameters of a given probability model. The improved computer capacity has rendered this method highly versatile in constructing phylogenetic trees. The ML method assesses the likelihood of the observed sequences based on the probability the chosen evolutionary model predicted. Compared to MP and ML approaches, Bayesian phylogenetic analysis is a more sophisticated approach to phylogenetics.

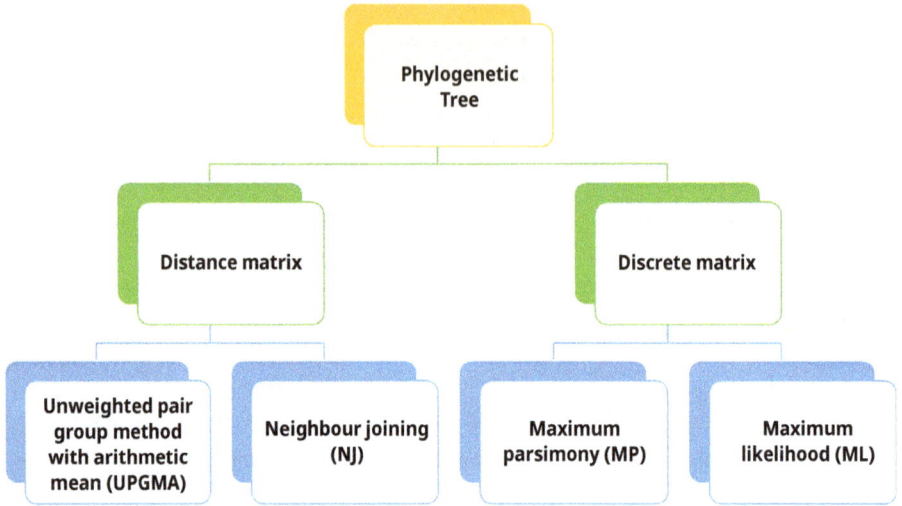

**Figure 8.2:** Classification of phylogenetic trees.

Since its introduction in the mid-1990s, the use of Bayesian statistics in phylogenetics has experienced significant advancements, leading to the widespread adoption of the Bayesian phylogenetic technique. The Bayesian approach involves making inferences about the probability of an unknown occurrence by calculating a posterior probability. In contrast with conventional statistical tests that merely depend on the available data to evaluate a hypothesis, Bayesian statistics include previous information alongside the present data to assess a hypothesis. Prior knowledge or available data can provide an initial estimate of the chance of an occurrence. However, by incorporating additional data and considering the primary probability, a more accurate estimation of the event's probability at a later time can be obtained. The application of established principles or empirical observations can demonstrate an initial understanding of probability. The Bayesian approach to tree construction involves iteratively sampling from a Monte Carlo Markov chain process. This technique aims to identify a tree topology that yields progressively higher scores as each sample is taken. The consensus tree with the highest posterior probability is then constructed using a group of highly scoring topologies for the tree. The Bayesian approach has several advantages over machine learning techniques, particularly in its ability to handle larger datasets and achieve quicker computational speeds effectively. MrBayes is an online Bayesian phylogenetic analysis program that may be used at the following URL: http://www.phylogeny.fr/one_task.cgi?task_type=mrbayes. In addition, the software can be obtained from the official website http://mrbayes.sourceforge.net/.

# 8.3 Various tools in phylogenetic analysis

Online phylogenetic tree-building tools are the most time-efficient option. Phylogeny.fr is an excellent online tool for conducting phylogenetic analysis, which may be accessed at http://www.phylogeny.fr/. The present server offers comprehensive phylogenetic analysis capabilities that serve individuals without specialized expertise. The construction of a phylogenetic tree can be improved by utilizing the One-Click option, which allows the user to generate the tree using the default values. MEGA (Molecular Evolutionary Genetics Analysis) version 5, developed by Tamura et al. [4], serves as another tool for generating phylogenetic trees [http://www.megasoftware.net/]. The MEGA software is user-friendly, with an easy-to-use user interface and self-explanatory toolbar.

Additionally, comprehensive instructions are available and elucidated by Hall [5]. Several commonly employed and adaptable software programs for conducting phylogenetic analysis include PHYLIP (Phylogeny Inference Package) (available at http://evolution.genetics.washington.edu/phylip.html), PAUP (Phylogenetic Analysis Using Parsimony), and MacClade (accessible at http://macclade.org/macclade.html). To obtain MacClade, users must navigate to the MacClade website and select "Acquiring MacClade" (http://macclade.org/download.html). There are various additional tools for phylogenetic analysis that may be accessed through the following website: http://molbiol-tools.ca/Phylogeny.html. Some of them are listed in Table 8.1. Overall, one can conclude that evolutionary relationships can be efficiently established through phylogenetic analysis, which has been instrumental in studying evolutionary biology, systematics, taxonomy, and modern drug discovery.

**Table 8.1:** List of online tools for phylogenetic analysis available on the web.

| S. no. | Online tools | Details | References |
|---|---|---|---|
| 1. | POWER (PhylOgenetic Web Repeater) | This software facilitates users in performing repeated phylogenetic analyses using several programs within the PHYLIP package. The system offers two distinct investigative pipelines: the multiple sequence alignment (MSA) pipeline and the aligned sequence phylogenetic analysis pipeline. It is relatively simple to handle. | [6] |
| 2. | RAxML (Randomized Axelerated Maximum Likelihood) | This software is employed for maximum likelihood sequence-based sequential and parallel inference of large phylogenetic trees. | [7] |

**Table 8.1** (continued)

| S. no. | Online tools | Details | References |
|---|---|---|---|
| 3. | Phylogeny.fr | This web tool is simple and is dedicated to reconstructing and analyzing phylogenetic relationships between molecular sequences. It is the combination of multiple alignments (MUSCLE, T-Coffee, ClustalW, and ProbCons), phylogeny (PhyML, MrBayes, TNT, and BioNJ), tree viewers (Drawgram, Drawtree, and ATV), and utility programs. | [8] |
| 4. | Phylemon2 | A suite of molecular evolution web tools, phylogenetics, and phylogenomics. | [9] |
| 5. | T-Rex (Tree and reticulogram REConstruction) | This methodology is employed to reconstruct phylogenetic trees and reticulation networks, as well as inferring occurrences of horizontal gene transfer (HGT). This program combines many bioinformatics techniques, including MUSCLE, MAFFT, neighbor joining, NINJA, BioNJ, PhyML, RAxML, a random phylogenetic tree generator, and established sequence-to-distance transformation models. | [10] |
| 6. | AmphoraNet | The online server implements the AMPHORA2 procedure to allocate a taxonomic category to each phylogenetic marker gene detected in the input metagenomic sample, accompanied by probability weighting. Metagenomic phylotyping utilizes a collection of 31 bacterial and 104 archaeal protein-encoding marker genes. The primary reason for the suitability of AmphoraNet in estimating the taxonomic makeup of bacterial and archaeal communities from metagenomic shotgun sequencing data lies in the abundance of single-copy genes within these communities. | [11] |

**Table 8.1** (continued)

| S. no. | Online tools | Details | References |
|---|---|---|---|
| 7. | FastME | This tool is based on an equilibrated minimum evolution with fast, sophisticated algorithms to perform topological moves. The first FastME version included only nearest neighbor interchange (NNI). The updated 2.0 version of the software incorporates subtree pruning and regrafting (SPR) alongside the existing NJ algorithm while maintaining its high computational efficiency. This enhanced version offers a range of features, including the ability to estimate distances between DNA and protein sequences using various models and settings, and support for bootstrapping and parallel computing. | [12] |
| 8. | CVTree3 | It was initially created to determine the evolutionary relationship between microorganisms, and it has since been effectively used to determine the relationships between viruses, chloroplasts, and fungi. A composition vector (CV) methodology enables the construction of phylogenetic trees based on whole-genome data without requiring sequence alignment. CVTree3 is a computational tool that facilitates the comparison of phylogenetic trees with taxonomic classifications. It assesses the monophyletic nature of tree branches spanning from the highest taxonomic rank (domain) to the lowest (species). | [13] |
| 9. | PhyML | Due to its ease of use and good coordination of speed and accuracy, it is widely utilized. | [14] |
| 10. | VICTOR (Virus Classification and Tree Building Online Resource) | This web-based tool facilitates the comparative analysis of viruses originating from bacteria and archaea by examining their genomic or protein sequences. The findings encompass phylogenetic trees derived using the GBDP (Genome-BLAST Distance Phylogeny) approach, branch support values, and recommendations for taxonomic classification at the species, group, and family levels. | [15] |

# References

[1]  Choudhuri S. Phylogenetic analysis. *Bioinformatics for Beginners*, 2014. 209–218. doi: 10.1016/b978-0-12-410471-6.00009-8.

[2]  Yang Z., Rannala B. Molecular phylogenetics: Principles and practice. *Nature Reviews Genetics*, 2012 May. **13**(5): 303–314.

[3]  Saitou N., Nei M. The neighbor-joining method: A new method for reconstructing phylogenetic trees. *Molecular Biology and Evolution*, 1987 Jul 1. **4**(4): 406–425.

[4]  Tamura K., Peterson D., Peterson N., Stecher G., Nei M., Kumar S. MEGA5: Molecular Evolutionary Genetics Analysis using maximum likelihood, evolutionary distance, and maximum parsimony methods. *Molecular Biology and Evolution*, 2011 Oct 1. **28**(10): 2731–2739.

[5]  Hall B.G. Building phylogenetic trees from molecular data with MEGA. *Molecular Biology and Evolution*, 2013 Mar 12. **30**(5): 1229–1235.

[6]  Lin C.Y., Lin F.K., Lin C.H., Lai L.W., Hsu H.J., Chen S.H., Hsiung C.A. POWER: PhylOgenetic Web Repeater – An integrated and user-optimized framework for biomolecular phylogenetic analysis. *Nucleic Acids Research*, 2005 Jul 1. **33**(suppl_2): W553–6.

[7]  Stamatakis A. RAxML-VI-HPC: Maximum likelihood-based phylogenetic analyses with thousands of taxa and mixed models. *Bioinformatics*, 2006 Nov 1. **22**(21): 2688–2690.

[8]  Dereeper A., Guignon V., Blanc G., Audic S., Buffet S. Phylogeny. fr: Robust phylogenetic analysis for the non-specialist nucleic acids research 36 (web server issue): W465-9. *Electronic Publication*, 2008.

[9]  Sánchez R., Serra F., Tárraga J., Medina I., Carbonell J., Pulido L., De María A., Capella-Gutíerrez S., Huerta-Cepas J., Gabaldón T., Dopazo J. Phylemon 2.0: A suite of web-tools for molecular evolution, phylogenetics, phylogenomics and hypotheses testing. *Nucleic Acids Research*, 2011 Jun 6. **39**(suppl_2): W470–4.

[10]  Boc A., Diallo A.B., Makarenkov V. T-REX: A web server for inferring, validating and visualizing phylogenetic trees and networks. *Nucleic Acids Research*, 2012 Jun 6. **40**(W1).

[11]  Kerepesi C., Bánky D., Grolmusz V. AmphoraNet: The webserver implementation of the AMPHORA2 metagenomic workflow suite. *Gene*, 2014 Jan 10. **533**(2): 538–540.

[12]  Lefort V., Desper R., Gascuel O. FastME 2.0: A comprehensive, accurate, and fast distance-based phylogeny inference program. *Molecular Biology and Evolution*, 2015 Oct 1. **32**(10): 2798–2800.

[13]  Zuo G., Hao B. CVTree3 web server for whole-genome-based and alignment-free prokaryotic phylogeny and taxonomy. *Genomics, Proteomics & Bioinformatics*, 2015 Oct 1. **13**(5): 321–331.

[14]  Lefort V., Longueville J.E., Gascuel O. SMS: Smart model selection in PhyML. *Molecular Biology and Evolution*, 2017 Sep 1. **34**(9): 2422–2424.

[15]  Meier-Kolthoff J.P., Göker M. VICTOR: Genome-based phylogeny and classification of prokaryotic viruses. *Bioinformatics*, 2017 Nov 1. **33**(21): 3396–3404.

**Multiple choice questions**

Q1  Which of the following is shortly defined as the evolutionary history of a kind of an organism?
a)  Recombinant pair
b)  Recessive tree
c)  Phylogeny
d)  Dominancy

Q2  In phylogeny, the phylogenetic tree is a two-dimensional graph that shows the evolutionary relationship between the _____.
a)  Organisms
b)  Species
c)  Genes
d)  All of above

Q3  The external node in the phylogenetic tree is also called as _____ and it represents the tip of the tree.
a)  Initial node
b)  Terminal node
c)  Synaptic node
d)  Permeable node

Q4  The _____ node is used to represent the LCA of the two lineage.
a)  Externodal
b)  External
c)  Parallel
d)  Internal

Q5  In unscaled branches, the length is _____ to the number of changes.
a)  Not parallel
b)  Not proportional
c)  Semiproportional
d)  Perpendicular

Q6  In _____ the branch length is proportional to the evolutionary change and is an example of scaled branch.
a)  Phylogram
b)  Cladogram
c)  Hologram
d)  Kilogram

Q7  While constructing phylogenetic trees, researchers identify _____ features that are shared by some species but not by others.
a)  Heterologous
b)  Different
c)  Homologous
d)  Heterozygous

Q8 The phylogenetic tree represents evolutionary relationship among species and also called as _____.
   a) Genetic tree
   b) Evolutionary tree
   c) Polycystic tree
   d) Polysaccharide tree

Q9 Which of the following is an essential preliminary to the tree reconstruction?
   a) Model build
   b) Model selection
   c) Sequence validation
   d) Sequence alignment

Q10 _____ is the derived character shared by more than one species or group.
   a) Synapomorphy
   b) Apomorphy
   c) Polymorphic
   d) None of them

---

**! Answers**

Q1  d)  All of above
Q2  d)  All of above
Q3  b)  Terminal node
Q4  d)  Internal
Q5  b)  Not proportional
Q6  a)  Phylogram
Q7  c)  Homologous
Q8  b)  Evolutionary tree
Q9  d)  Sequence alignment
Q10 a)  Synapomorphy

Sheetal Dagar, Anil Panwar*, Varruchi Sharma, Imran Sheikh,
Vikas Kushwaha, Damanjeet Kaur, Anil K. Sharma, and Sri Kant

# Chapter 9
# Basic structure of proteins: current paradigms, trends, and perspective

**Abstract:** Proteins are essential biomolecules, serving as the building blocks of living organisms and playing vital roles in various biological processes. Amino acids, the fundamental units of proteins, consist of an amino group, a carboxyl group, and a unique side chain (R group). There are around 500 known amino acids, with 20 forming the genetic code. The structure of proteins can be categorized into primary, secondary, tertiary, and quaternary levels. Primary structure represents the linear sequence of amino acids, while the secondary structure involves the folding of segments into alpha helices and beta sheets. Tertiary structure results from long-range interactions between amino acids, and quaternary structure refers to the arrangement of multiple polypeptide chains. Protein motifs and domains are specific arrangements of secondary structures, providing functional and structural characteristics. Techniques such as nuclear magnetic resonance (NMR) spectroscopy and X-ray crystallography allow the determination of protein 3D structures. NMR is particularly valuable in native and dynamic environments, while X-ray crystallography requires crystallization. Understanding protein structures aids in drug design, biotechnology, and various research fields, contributing to advancements in science and medicine. However, challenges, such as obtaining crystals and nonphysiological conditions, remain in protein structure determination.

**Keywords:** Biomolecules, protein motifs, NMR spectroscopy, X-ray crystallography

*Corresponding author: Anil Panwar, Department of Bioinformatics and Computational Biology, CCS Haryana Agricultural University, Hisar 125004, Haryana, India, e-mail: kumarniki2003@gmail.com
Sheetal Dagar, Department of Bioinformatics and Computational Biology, CCS Haryana Agricultural University, Hisar 125004, India
Varruchi Sharma, Vikas Kushwaha, Damanjeet Kaur, Department of Biotechnology and Bioinformatics, Sri Guru Gobind Singh College, Sector 26, 160019 Chandigarh
Imran Sheikh, Department of Biotechnology, Eternal University, Baru Sahib, Sirmour, Himachal Pradesh, India
Anil K. Sharma, Department of Biotechnology, Amity University, Sector 82 A, IT City Road, Block D, Sahibzada Ajit Singh Nagar, Mohali 140306, Punjab, India
Sri Kant, Academic Affairs Department, Chandigarh University, Gharuan, Mohali 140413, Punjab, India

https://doi.org/10.1515/9783111568584-009

## 9.1 Amino acids

Amino acids are building blocks of proteins [1]. Amino acids are a class of organic compounds having two functional groups, amine ($-NH_2$) and carboxyl ($-COOH$), and a side chain (R) that is unique for each amino acid [2]. Proteins comprise 20 different types of amino acids, and each side chain (R group) varies according to the chemical structures they contain (Figure 9.1). The R group represents the unique side chains responsible for the chemical properties of individual amino acids [3]. These characteristics include size, shape, interactions, hydrophilicity, hydrophobicity, polarity, and pH level. The stability of proteins in the body and environment depends on each of these qualities [4]. The amino ($-NH_2$) group is basic, whereas the carboxyl ($-COOH$) group is acidic. Amino acids are chemical molecules made up of specific R (side chain) groups, hydrogen (H), amino ($-NH_2$), carboxyl ($-COOH$), and an alpha carbon in the middle [4]. The term "amino acid" is an abbreviation for α-amino acid. α-Amino acids are unique because the amino and carboxylic acid functional groups are separated by only one carbon atom, which is usually a chiral carbon [5]. The remaining two bonds of the α-carbon atom are usually filled by hydrogen atoms (H) and R groups. Since the rest of the structure is the same, the property of these amino acids is mainly determined by side chain groups. More than 300 amino acids have been described, but only 20 amino acids take part in protein synthesis [6].

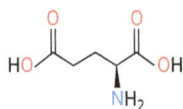

**Figure 9.1:** Structure of a typical amino acid.

### 9.1.1 Classification

Amino acids can be classified into the following five types:
1. According to the structure
2. Based on the $-NH_2$ position
3. Based on nutritional requirements
4. Based on metabolic fates
5. According to the polarity

#### 9.1.1.1 Classification according to the structure

1. **Amino acids with aliphatic side chains:** They are known as monoamine monocarboxylic acids. This group includes the five most basic amino acids: glycine, alanine, valine, leucine, and isoleucine. Because the last three amino acids (leucine,

isoleucine, and valine) have branched aliphatic side chains, they are called branched-chain amino acids.

2. **Amino acids with a hydroxyl group:** The amino acids serine, threonine, and tyrosine all have hydroxyl groups. Tyrosine is usually categorized as an aromatic amino acid due to its aromatic nature.

3. **Amino acids with sulfur groups:** Methionine with a thioether group and cysteine with a sulfhydryl group are the two amino acids that are ingested during protein synthesis.

4. **Acidic amino acids:** Asparagine and glutamine are the respective amide derivatives of aspartic acid and glutamic acid, which are dicarboxylic monoamine acids.

5. **Basic nature of amino acids:** The three dibasic monocarboxylic acids are lysine, arginine (with a guanidino group), and histidine (with an imidazole ring). They have very fundamental properties.

6. **Amino acids have an aromatic ring:** Tryptophan, tyrosine, and phenylalanine are examples of aromatic amino acids (which have an indole ring). Histidine may potentially fall under this group in addition to them.

   Proline, an amino acid with a pyrrolidine ring, is an uncommon amino acid. It has an imino group (=NH) rather than the amino group ($-NH_2$) present in other amino acids.

7. **Heterocyclic amino acids:** Proline, histidine, and tryptophan.

### 9.1.1.2 Classification according to the $-NH_2$ position

Amino acids can be divided into three types:

1. α-Amino acids: An amino group attached to the carbon atom next to the carboxyl group is called an α-amino acid. All naturally occurring amino acids are found as α-L-amino acids.

2. β-Amino acids: A β-amino acid refers to an amino group attached to the third carbon atom.

   Amino acids (numbering from the carboxyl group).

3. γ-Amino acids: A γ-amino acid refers to the amino group attached to the fourth carbon atom of the amino acid.

   Carbon (numbering from the carboxyl group): For example, GABA (γ-aminobutyric acid).

### 9.1.1.3 Classification based on nutritional needs

Based on nutritional requirements, amino acids can be classified into three types:
1.  **Essential amino acids (EAA):** Some amino acids cannot be synthesized in the human body and must be obtained from food. This is essential for the proper development and maintenance of the individual.
2.  **Non-EAA:** The body can synthesize about 10 amino acids to meet its biological needs, so it does not need them; for example, glycine, alanine, serine, cysteine, asparagine, aspartic acid, glutamic acid, glutamine, tyrosine, and proline.
3.  **Semi-essential amino acids:** The amino acids histidine and arginine are semi-essential. They are necessary for the growth and nutrition of children. However, mature individuals do not need them.

### 9.1.1.4 Classification based on the metabolic fate

Amino acids can be divided into three groups based on their metabolic fate.
1.  **Pure ketogenic amino acids**: These amino acids can be used to make fats since leucine becomes ketone bodies. L-Lysine is fully ketogenic.
2.  **Ketogenic and glucogenic amino acids**: On the other hand, some of the carbon skeletons of these amino acids enter the ketogenic pathway, while others enter the glucogenic pathway. For example, lysine, isoleucine, phenylalanine, tyrosine, and tryptophan are partly ketogenic and glucogenic.
3.  **Pure glucogenic amino acid**: These amino acids can be used as starting points to produce glucose or glycogen. All the remaining 14 amino acids are intact as they only enter the glucogenic pathway.

### 9.1.1.5 Classification by polarity

Amino acids can be divided into four groups based on their polarity (Table 9.1).
1.  **Nonpolar amino acids:** These amino acids, also known as hydrophobic (water-repellent) amino acids, carry no electrical charge in the "R" group; for example, alanine, leucine, isoleucine, valine, methionine, phenylalanine, tryptophan, and proline.
2.  **Polar amino acids with no charge on the "R" group:** As a result, the "R" groups of these amino acids are uncharged. However, there are groups like hydroxyls, sulfhydryls, and amides that contribute to hydrogen bonding in protein structure; for example, glycine, serine, threonine, cysteine, glutamine, asparagine, tyrosine.
3.  **Polar amino acids with positive "R" groups:** For example, lysine, arginine, and histidine.
4.  **Polar amino acids with negative "R" groups:** Aromatic dicarboxylic monoamine acids, for example, aspartic acid and glutamic acid.

**Table 9.1:** Twenty amino acids with their abbreviations and symbols.

| Amino acid | Three-letter abbreviation | One-letter symbol |
|---|---|---|
| Alanine | Ala | A |
| Arginine | Arg | R |
| Asparagine | Asn | N |
| Aspartic acid | Asp | D |
| Cysteine | Cys | C |
| Glutamine | Gln | Q |
| Glutamic acid | Glu | E |
| Glycine | Gly | G |
| Histidine | His | H |
| Isoleucine | Ile | I |
| Leucine | Leu | L |
| Lysine | Lys | K |
| Methionine | Met | M |
| Phenylalanine | Phe | F |
| Proline | Pro | P |
| Serine | Ser | S |
| Threonine | The | T |
| Tryptophan | Trp | W |
| Tyrosine | Tyr | Y |
| Valine | Val | V |

## 9.2 Peptides

A peptide is a molecule composed of two or more amino acids linked together by peptide bonds. Peptide bonds are covalent chemical bonds formed through a condensation reaction, which involves the removal of a water molecule. Peptides can vary in length from very short chains containing just a couple of amino acids to longer chains with up to around 50 amino acids [7]. A peptide bond is defined as a covalent bond between the carbonyl group of one amino acid and the amino group of another amino acid. This bond is also called an amide bond. The end of a peptide chain that has a free amino group ($NH_2$) is referred to as the amino-terminal or N-terminal residue. Conversely, the end with a free carboxyl group (COOH) is called the carboxyl-terminal or C-terminal residue. These N-terminal and C-terminal residues are important in defining the directionality of the peptide chain and play a role in the peptide's interactions and functions within biological systems [2]. The amino terminus is usually placed on the left and the carboxy terminus on the right. Therefore, the sequence is read from left to right (Figure 9.2).

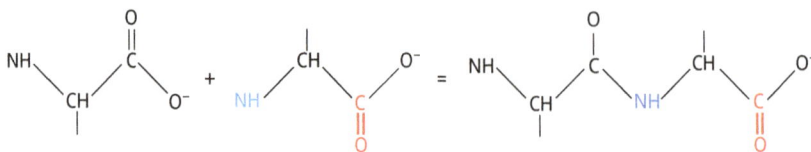

Figure 9.2: Synthesis of a peptide bond.

## 9.3 Dihedral angles

The hydrogen atom of the amide nitrogen and the oxygen atom of the carbonyl group are trans to one another, and the six peptide group atoms are arranged in a single plane. Pauling and Corey deduced from their data that the partial double-bond nature of the peptide CON bonds prevents them from rotating freely. Rotation around the NOC and COC bonds is allowed. Thus, the backbone of a polypeptide chain can be visualized as a collection of rigid planes that all have the same point of rotation, C, in common. A variety of conformations that a polypeptide chain can acquire are constrained by the stiff peptide bonds.

Conventionally, the NOC bond's bond angle is denoted by $\emptyset$ (phi) while the COC bond's bond angle is denoted by $\varphi$ (psi). When the polypeptide is in its completely expanded shape and all the peptide groups are in the same plane, $\varphi$ and $\emptyset$ are defined as 180. For this reason, the conformation where both angles are 0 is forbidden; the angles of rotation are only described in relation to this conformation. When plotted against each other in a Ramachandran plot introduced by G. N. Ramachandran, the permitted values for these angles are visually displayed.

## 9.4 Proteins

Protein is the primary structural and functional element of every cell in the body. The essential amino nitrogen group is what distinguishes an amino acid or protein. An amino acid can be distinguished from sugar by its nitrogen group. The macromolecules known as proteins are made up of lengthy chains of amino acid subunits. Peptide bonds hold the amino acids together in the protein molecule. The chains that are generated in biological systems can range in size from a few amino acids (di-, tri-, or oligopeptide) to hundreds of units (polypeptide). The primary structure of the chain is the arrangement of the amino acids. The complexity of proteins' physical structures is a key characteristic.

Instead of existing as long chains, polypeptide chains fold into a three-dimensional (3D) structure. Amino acid chains are prone to helical bending (secondary structure). Hydrophobic interactions between non-polar side chains and, in the case of some proteins, disulfide bonds can cause sections of the helices to fold onto each other, giving the whole molecule a globular or rod-like shape (tertiary structure). Their exact shape depends on their function and, for some proteins, on interactions with other molecules (quaternary structure).

The most adaptable macromolecules in living systems, proteins are crucial to almost all biological activities. They serve as catalysts, move and store other molecules like oxygen, offer immunological protection and mechanical support, produce movement, send nerve impulses, and regulate development and differentiation. All proteins are made up of the same universal set of 20 amino acids that are covalently bonded to create linear sequences, whether they come from the most primitive types of bacteria or the most sophisticated forms of life. Each of these amino acids has a side chain with unique chemical characteristics. Proteins with very diverse characteristics may be produced by cells in a wide variety of combinations and sequences. The specific 3D structure and function of the protein are determined by the amino acid sequence.

The categorization of proteins encompasses two primary types: simple proteins and complex proteins. A simple protein exclusively comprises amino acid residues, whereas a complex protein integrates both amino acid and non-amino acid constituents. The non-amino acid component within a complex protein is denoted as a "prosthetic group." For example, glycoproteins contain carbohydrates. Similarly, lipoproteins and metalloproteins contain lipids and specific prosthetic metal groups, hence the name metalloproteins. Few proteins have multiple prosthetic groups.

## 9.5 Hierarchy of proteins

A protein can only perform as intended when it is in the proper conformation or 3D shape. Here, we focus on the four layers of organization that make up proteins, starting with their monomeric building components, the amino acids.

## 9.6 Primary structure

The term "primary structure" initially referred to a protein's whole covalent structure, but it is now more commonly used to refer to the order of amino acids in each polypeptide chain that makes up the protein. The linear arrangement, or sequence, of the amino acid residues that make up a protein constitutes its basic structure. The chains created when amino acids polymerize are referred to by various names. A peptide is a small chain of amino acids with a specific sequence connected by peptide bonds; larger chains are known as polypeptides. In contrast to polypeptides, which may have up to 4,000 residues, peptides typically have 20–30 amino acid residues or less. The name "protein" is typically reserved for polypeptides (or complexes of poly-

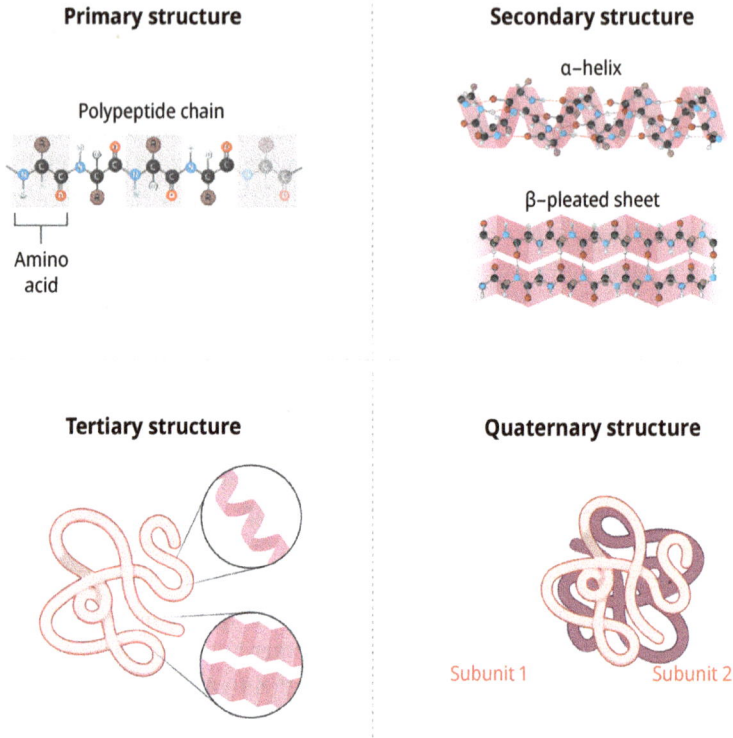

**Figure 9.3:** Hierarchy of proteins.

peptides) with well-defined 3D structures (Figure 9.3). Proteins and peptides are as-sumed to be the by-products of a cell. Proteins and peptides are regarded as cellular by-products. The dimension of a protein or polypeptide is quantified in relation to its mass, denominated in dalton (Da), an atomic mass unit. Alternatively, the measurement is expressed as molecular weight (MW), a unitless numerical value. For instance, a pro-tein with an MW of 10,000 translates to a mass of 10,000 Da or equivalently 10 kDa [8].

## 9.7 Secondary structure

Secondary structures are different spatial configurations formed by folding localized segments of the polypeptide chain at the second level of the protein structure hierar-chy. Depending on its sequence, a polypeptide can exhibit different secondary struc-tures. The polypeptide adopts a random coil structure in the absence of stabilizing noncovalent interactions. However, segments of the protein backbone adopt specific recurring configurations, known as secondary structures. These structures include the alpha helix, beta sheet, and compact U-turn, which are formed due to the stabilization of hydrogen bonds between specific amino acid residues.

## 9.8 Alpha helix

The carbonyl oxygen atom of each peptide bond is hydrogen bonded to the amino acids and the amide hydrogen atom four residues toward the C-terminus in a polypeptide segment that has been folded into a helix. As all the hydrogen bond donors have the same orientation, this periodic arrangement of bonds gives the helix directionality. An example is the fibrous protein keratin, which is organized in this manner, and many globular proteins also incorporate helical segments in their structure (Figure 9.4).

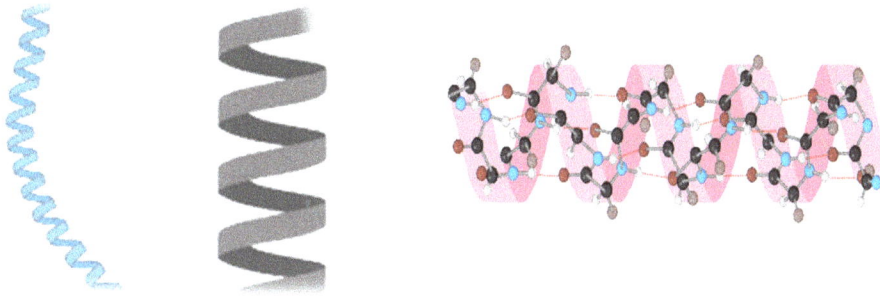

**Figure 9.4:** Alpha helix.

## 9.9 Beta sheet

The sheet is a different kind of secondary structure made up of strands that are packed laterally. A brief (five to eight residues), almost completely expanded polypeptide segment makes up each of the strands. A sheet is created by the formation of hydrogen bonds between backbone atoms in neighboring strands, either within the same polypeptide chain or between distinct polypeptide chains. The hydrogen bonds between the main-chain carbonyl (CO) and amino (NH) groups occur between adjacent strands of the sheet. The orientation of the peptide bond determines the directionality of strands, much like it does for helices. In a parallel β-sheet arrangement, all the strands that form the sheet run parallel to each other and follow the same direction from N to C termini. Conversely, an antiparallel β-sheet is formed when the strands remain parallel to each other, but they run in opposite directions [9] (Figure 9.5).

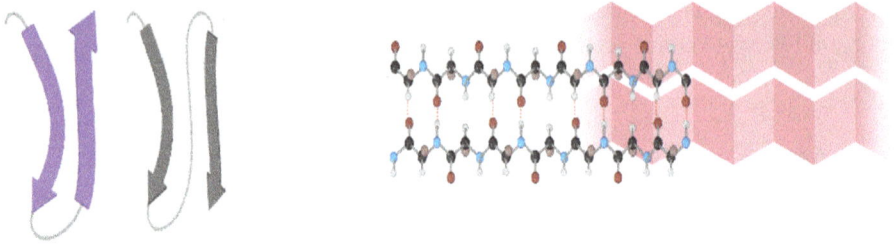

**Figure 9.5:** Beta sheets in parallel and antiparallel directions.

## 9.10 Beta turns

Turns are abrupt bends on the surface of a protein that drive the polypeptide backbone back to the inside. Turns are composed of three or four amino acids. A hydrogen bond between the end residues of these brief, U-shaped secondary structures serves to stabilize them. Glycine and proline are frequently found together in beta turns. The polypeptide backbone may fold into a tight U shape since proline has a built-in bend and glycine does not have a significant side chain. Turns enable the folding of large proteins into very compact structures. Longer bends or loops are also possible in polypeptide backbones. Loops may be created in a variety of ways, unlike turns, which have a limited number of clearly defined patterns.

## 9.11 Loops or coils

Along with the secondary structures already mentioned, polypeptide chains include a substantial quantity of unordered structures in which the psi and phi angles are not identical. The conformation of these unordered structures is a coil or loop.

## 9.12 Tertiary structure

Tertiary structure is formed as a result of secondary structure going through tertiary folding (Figure 9.6). Long-range interactions, including hydrogen bonds, disulfide cross-links, hydrophobic contacts, electrostatic interactions, and van der Waals interactions, help to maintain the tertiary structure. Different forms of secondary structure may interact to generate tertiary structure in this type of structure, which brings amino acids that are far from one another in the polypeptide sequence close together. Nuclear magnetic resonance (NMR) spectroscopy and X-ray diffraction are two methods that may be used to examine protein's tertiary structures. In the tertiary structure, polar residues are found on the protein's surface, whereas hydrophobic amino acids are hidden inside the molecule.

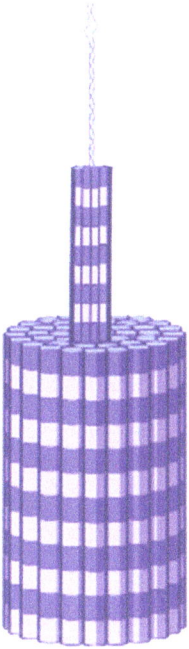

**Figure 9.6:** Structure of collagen.

# 9.13 Motifs and domains

A protein's tertiary structure is composed of specific arrangements of secondary structures known as motifs or folds. Protein motifs are discrete regions with similar amino acid sequences or 3D protein structures. They are recognized regions of the protein structure that may or may not have a specific chemical or biological activity. Motifs can serve as the distinctive features of a certain function in various circumstances. One $Ca^{2+}$-binding pattern is the helix–loop–helix, which is identified by certain hydrophilic residues present at invariant places in the loop. Another typical motif is the zinc finger, which consists of three secondary structures: a helix, two antiparallel strands, and a finger-like bundle bound together by a zinc ion (Figure 9.7). A structural motif is like a fold, but it is typically smaller in size and often serves as the foundational unit for larger folds. Certain structural motifs, such as β-barrels, are found in a wide range of unrelated proteins with numerous variations. On the other hand, some motifs are particularly conserved due to their connection to specific biochemical functions, like zinc fingers in DNA-binding motifs [9]. The stable and often independently folding part of the overall protein structure is known as the "structural domain." A structural domain is made up of 100–150 residues arranged in different patterns.

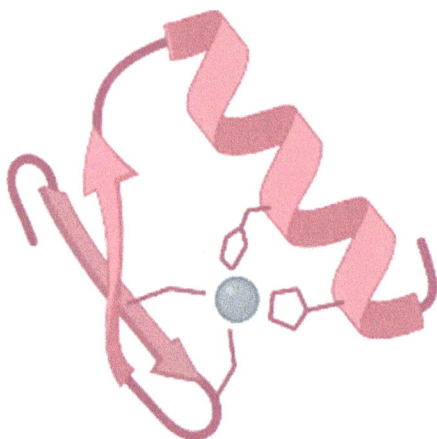

**Figure 9.7:** Zinc finger motif.

## 9.14 Domain

The term "domain" is employed to denote the fundamental units of protein structure (tertiary) and function. Typically, a polypeptide comprising 200 amino acids is composed of two or more domains [10]. A domain is frequently distinguished by an intriguing structural characteristic, such as an unusually high composition of specific amino acid sequences that are shared by (conserved in) many proteins, or a specific secondary structure motif (e.g., the zinc finger motif in the kringle domain).

Based on observations showing that a protein's activity is localized to a certain area along its length, domains are occasionally described in terms of their functional properties. For example, a specific section or parts of a protein may oversee that protein's catalytic activity (e.g., a kinase domain) or binding capacity (e.g., a DNA-binding domain and a membrane-binding domain).

## 9.15 Quaternary structure

The configuration of numerous proteins, particularly those surpassing a mass of 100 kDa, exists as oligomers comprising multiple polypeptide chains. The exact spatial organization of these subunits within a protein is denoted as its quaternary structure [9]. These polypeptide chains or subunits might be alike or dissimilar, resulting in homogeneous or heterogeneous quaternary structures, respectively. Since the hemoglobin proteins are composed of two alpha-chains and two beta-chains, they have a heterogeneous quaternary structure. Aspartate transcarbamylase is similar in that it has 12 polypeptide chains or subunits, 6 catalytic and 6 regulatory, arranged in 2 catalytic trimers and 3 regulatory dimers. Proteins are classified as fibrous or globular based on their greater level of structure.

# 9.16 Methods for determination of proteins' 3D structure

Protein structures may be determined using NMR spectroscopy, which has become a significant alternative to X-ray crystallography since it resembles physiological circumstances as closely as possible (i.e., in solution).

**NMR spectroscopy** is a common technique for determining the atomic-level structures of proteins and various protein complexes. It can also provide detailed information about the interactional and conformational dynamics that take place in a variety of sample states, from diluted solutions to living cells, and that occur over a range from picoseconds to seconds or even days (Figure 9.8).

Protein NMR structure determination procedures typically involve the following four steps:
1) Preparing isotope-labeled protein samples
2) Collecting and analyzing NMR data, and specifically determining the chemical shifts of the 1H, 15N, and 13C atoms in the protein molecule
3) Employing distance and/or orientation constraints acquired from NMR data, structure computation, and refinement
4) Structural quality evaluation, each of which is presented and examined in detail [11]

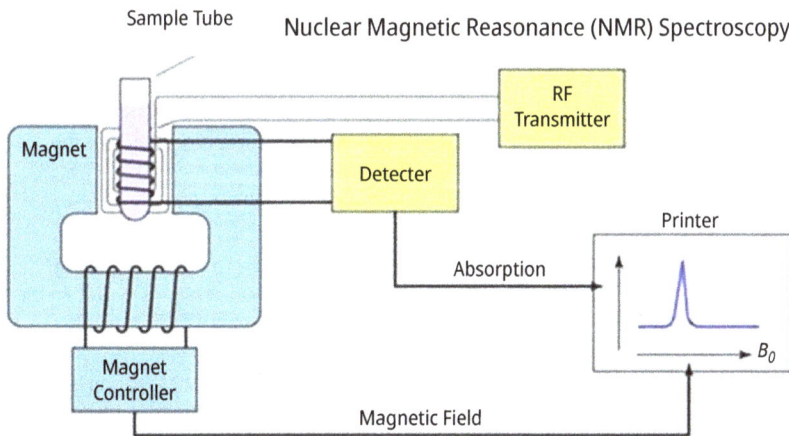

**Figure 9.8:** NMR spectroscopy instrumentation.

NMR spectroscopy has unraveled numerous enigmas regarding the behaviors of molecules. Protein NMR spectroscopy is instrumental in examining alterations in protein conformation, denaturation, and internal mobility, as well as pH-induced changes in ionizable amino acid side chains within enzyme's active sites. It enables the observation of hydrogen-bonded imino protons in tRNA, exploration of paramagnetic centers in metalloproteins, and more. This technique gauges the distances and angles between various protons, which are then utilized computationally to deduce the protein's structure. Contemporary protein spectroscopy necessitates multidimensional experiments, encompassing 1H, 13C, and 15N nuclei within isotopically labeled proteins [12] (Figure 9.9).

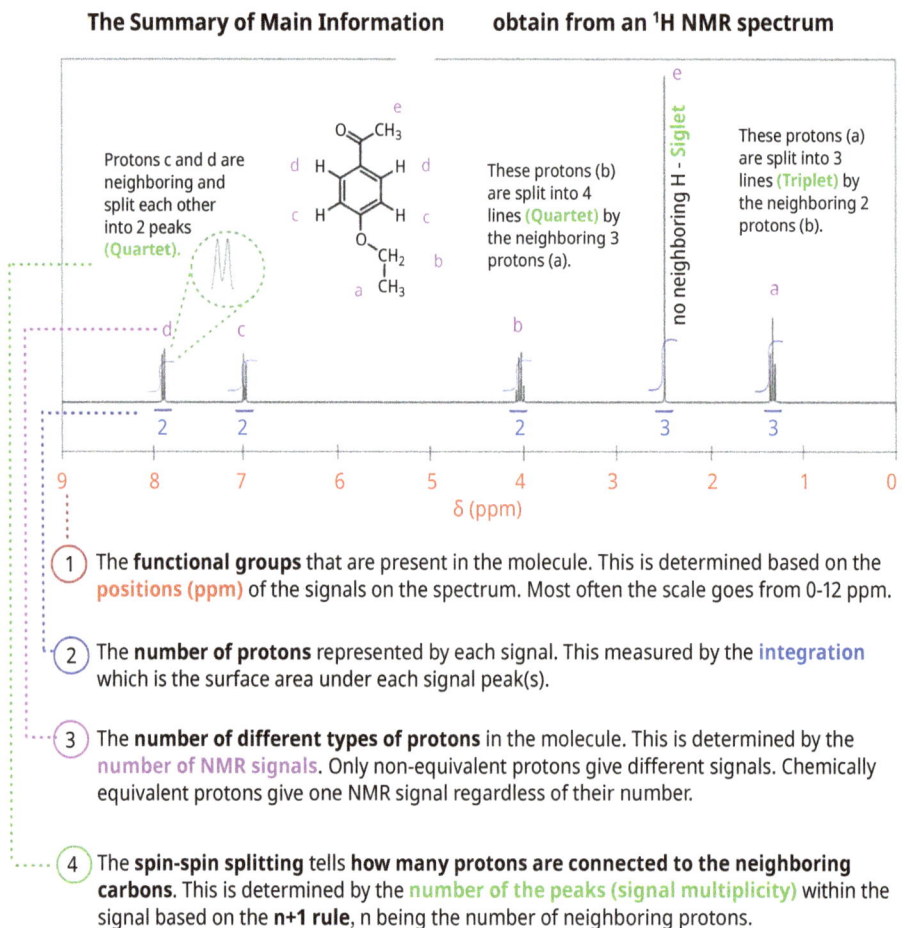

**The Summary of Main Information     obtain from an ¹H NMR spectrum**

1. The **functional groups** that are present in the molecule. This is determined based on the positions (ppm) of the signals on the spectrum. Most often the scale goes from 0-12 ppm.

2. The **number of protons** represented by each signal. This measured by the integration which is the surface area under each signal peak(s).

3. The **number of different types of protons** in the molecule. This is determined by the number of NMR signals. Only non-equivalent protons give different signals. Chemically equivalent protons give one NMR signal regardless of their number.

4. The **spin-spin splitting** tells **how many protons are connected to the neighboring carbons**. This is determined by the number of the peaks (signal multiplicity) within the signal based on the **n+1 rule**, n being the number of neighboring protons.

**Figure 9.9:** An example of the data from an NMR machine. This two-dimensional NMR spectrum is derived from the C-terminal domain of the enzyme cellulase. The spots represent interactions between hydrogen atoms that are near neighbors in the protein and hence reflect the distance that separates them. Complex computing methods, in conjunction with the known amino acid sequence, enable possible compatible structures to be derived.

# 9.17 X-ray crystallography

X-ray crystallography is a technique for determining a crystal's atomic and molecular structure. The essential idea is that the crystalline atoms cause an X-ray beam to diffract in a variety of directions. A crystallographer may create a 3D representation of the density of electrons within the crystal by measuring the angles and intensities of these diffracted beams. The mean locations of the atoms in the crystal, as well as their chemical bonds, disorder, and other information, may be calculated using this electron density picture. Many biological compounds, including vitamins, medicines, proteins, and nucleic acids such as DNA, have their structure and function exposed by this approach (Figure 9.10).

Seven key stages are involved in determining the crystal structure of a protein:
1. Protein synthesis and purification
2. The formation of crystals
3. Characterization of crystals
4. Data collection and analysis
5. Phasing
6. Refinement
7. Structure analysis, validation, and deposition are all parts of the process [13]

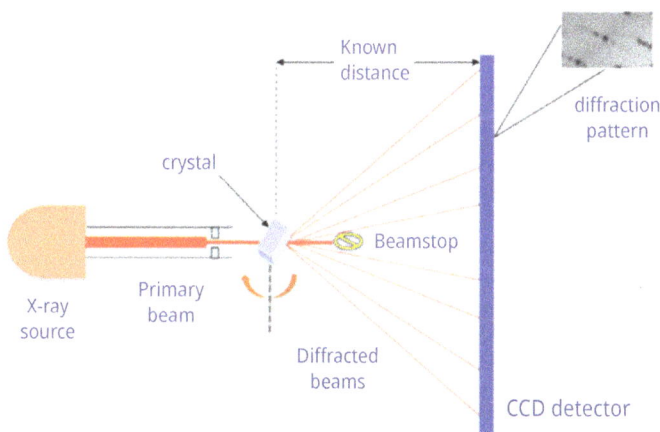

**Figure 9.10:** Schematic representation of an X-ray diffraction experiment.

About 85% of all known protein structures have been solved using X-ray crystallography. Knowledge of the 3D structure of proteins is useful in biotechnology, biomedicine, as a tool for drug design and basic research, and for the validation of protein modification, ligand binding, and structural integrity. In addition, the requirement for pure, homogeneous, and stable protein solutions in crystallization makes X-ray crystallography beneficial in other areas of protein research.

The advancement of computational technology in recent years, as well as the creation of new and powerful computer programs, together with the massive increase in the number of protein structures stored in the Protein Data Bank (PDB), makes the resolution of novel structures simpler than in the past. This approach enables the resolution of macromolecular structures using solely computer effort, starting with a single dataset and a search model acquired from the PDB.

The main limitation of X-ray crystallography is the difficulty in obtaining a crystal of viral particles, which is required for X-ray crystallography. Another limitation is that X-ray crystallography often necessitates the placement of samples in nonphysiological conditions, which can occasionally result in functionally irrelevant conformational changes.

# 9.18 Protein secondary and tertiary structure prediction (programs)

## Homology modeling

| Tool/software | Algorithms |
| --- | --- |
| MODELLER | Satisfaction of spatial restraints and comparative modeling |
| SWISS-MODEL | Template-based modeling and comparative modeling |
| Phyre2 | Profile–profile comparison and hidden Markov models |
| I-TASSER | Iterative Threading ASSEmbly Refinement and Fragment Assembly |
| RaptorX | Deep learning and template-based modeling |
| HHpred | Profile hidden Markov models |
| Prime | Comparative modeling and ab initio loop prediction |
| ModLoop | Ab initio loop prediction |
| QUARK | Ab initio modeling |
| Robetta | Ab initio folding and comparative modeling |
| SWISS-MODEL Workspace | Template-based modeling and comparative modeling |
| PhyreRisk | Comparative modeling and structural variation analysis |

## Threading/fold recognition

| Tool/software | Algorithms |
| --- | --- |
| HHpred | Profile hidden Markov models (HMMs) |
| FFAS | Profile–profile comparison |
| Phyre2 | Profile–profile comparison and HMMs |
| FUGUE | Profile–profile comparison |
| MUSTER | Profile–profile comparison |
| pGenTHREADER | Profile–profile comparison and HMMs |

(continued)

| Tool/software | Algorithms |
|---|---|
| **RaptorX** | Deep learning and template-based methods |
| **GeneSilico MetaServer** | Various methods, including HMMs and secondary structure |
| **3D-PSSM** | Position-specific scoring matrix |
| **SAM-T99/T98** | Sequence-structure alignment methods |
| **3D-BLAST** | Sequence-structure alignment methods |
| **FOLDpro** | Profile–profile comparison and threading |

## Ab initio structure prediction

| Tool/software | Algorithms |
|---|---|
| **Rosetta** | Monte Carlo and molecular mechanics |
| **I-TASSER** | Replica exchange Monte Carlo and molecular dynamics |
| **CABS-fold** | Coarse-grained Monte Carlo |
| **QUARK** | Iterative threading assembly refinement |
| **ROSETTA3D** | Monte Carlo with fragment assembly |
| **CONFOLD** | Metropolis Monte Carlo and simulated annealing |
| **TASSER** | Monte Carlo |
| **PEP-FOLD3** | Fragment assembly and simulated annealing |
| **FALCON** | Fragment assembly |
| **UNRES** | Coarse-grained Monte Carlo |
| **Modeller** | Comparative modeling |
| **RAPTORX-ALPHA** | Deep learning |
| **FRAGFOLD** | Fragment assembly and replica exchange Monte Carlo |

## Secondary structure prediction

| Tool/software | Method/algorithm |
|---|---|
| PSIPRED | Neural networks |
| JPred | Hidden Markov models |
| SOPMA | Neural networks |
| DSSP | Dictionary-based method |
| HHpred | Profile hidden Markov models |
| Spider2 | Deep learning |
| PSSpred | Position-specific scoring matrix |
| SCRATCH | Neural networks |
| PHYRE2 | Profile-based approach |
| SSpro | Multiple methods, including neural networks and hidden Markov models |

## 9.19 Conclusion

In conclusion, proteins are essential molecules in living organisms, serving a wide range of functions crucial for their survival. Proteins are composed of amino acids, which are the building blocks of the protein structure. Amino acids have different properties based on their side chains, which determine the characteristics of each protein. The structure of proteins can be classified into four hierarchical levels: primary, secondary, tertiary, and quaternary structures. The primary structure refers to the linear sequence of amino acids in the polypeptide chain. Secondary structure involves the folding of localized segments of the polypeptide chain into alpha helices, beta sheets, and turns. Tertiary structure results from long-range interactions between amino acids that bring distant segments of the polypeptide chain close together, forming a 3D conformation. Specific arrangements of secondary structures form motifs or folds, while the combination of different polypeptide chains results in the quaternary structure. Determining the 3D structure of proteins is crucial for understanding their function and drug design. Techniques like X-ray crystallography and NMR spectroscopy are commonly used to determine protein structures. Additionally, computational methods have been developed for protein structure prediction, which is essential when experimental techniques are challenging. Understanding the protein structure and function is fundamental for advancing various fields, including biotechnology, drug development, and basic research. In conclusion, the study of protein structure and function continues to be a fascinating and essential area of research, providing insights into the molecular basis of life and offering opportunities for the development of new therapeutic approaches and biotechnological applications.

## References

[1]  Berg J.M., Tymoczko J.L., Stryer L. Biochemistry, New York: W. H. Freeman; 4th Edn, 2002, 86.

[2]  Cox M.M., Nelson D.L. Lehninger Principles of Biochemistry 2013, 6th Edn, WH Freeman and Company; New York (USA.

[3]  Dietzen D.J. Principles and Applications of Molecular Diagnostics. *Amino Acids, Peptides, and Proteins*, 2018. **346**: 345–380.

[4]  Philip G.K., Freeland S.J. Did evolution select a non-random "alphabet" of amino acids? *Astrobiology*, 2011. **11**(3): 235–240.

[5]  Wu G. Amino acids: Metabolism, functions, and nutrition. *Amino Acids*, 2009. **37**(1): 1–17.

[6]  Akram M., Asif H.M., Uzair M., Akhtar N., Madni A., Shah S.A., . . . Ullah A. Amino acids: A review article. *Journal of Medicinal Plants Research*, 2011. **5**(17): 3997–4000.

[7]  Friedberg F., Winnick T., Greenberg D.M. Peptide synthesis in vivo. *The Journal of Biological Chemistry*, 1947. **169**(3): 763.

[8]  Alberts B., Johnson A., Lewis J., et al. Molecular Biology of the Cell, 4th Edn, New York: Garland Science; 2002. Analyzing Protein Structure and Function.

[9]    Sun P.D., Foster C.E., Boyington J.C. Overview of protein structural and functional folds. *Current Protocols in Protein Science, Chapter*. **17**(1): 1711–17.1.

[10]   Satyanarayana U., Chakrapani U. Biochemistry, 4th Edn, Books & Allied Pvt. Ltd; 2013, 44–52.

[11]   Patel G., Chudasama D. *Nuclear Magnetic Resonance Spectroscopy NMR*, 2022. **11**: 30–34.

[12]   Kurt W. Protein structure determination in solution by NMR spectroscopy. *The Journal of Biological Chemistry*, 1990. **265**(36): 22059–22062.

[13]   Papageorgiou A.C., Mattsson J. Protein structure validation and analysis with X-ray crystallography. *Methods in Molecular Biology (Clifton, N.J.)*, 2014. **1129**: 397–421.

[14]   Sun P.D., Foster C.E., Boyington J.C. Overview of protein structural and functional folds. *Current Protocols in Protein Science, Chapter*. **17**(1): 1711–17.1.

[15]   Wade L.G. Jr. Organic chemistry. *W. Preston Reeves Journal of Chemical Education*, 1988. **65**(6): 1189.

## Multiple choice questions

Q1    _____ is the primary structure of a protein.
   a)   The sequence of amino acids
   b)   The overall 3D shape of the protein
   c)   The arrangement of multiple protein subunits
   d)   The presence of disulfide bonds between cysteine residues

Q2    _____ are the bonds responsible for stabilizing the secondary structure of proteins.
   a)   Peptide bonds
   b)   Hydrogen bonds
   c)   Ionic bonds
   d)   Disulfide bonds

Q3    The basic characteristic structure of an alpha helix is _____.
   a)   A flat, sheet-like structure
   b)   A coiled structure stabilized by hydrogen bonds between amino acids
   c)   A structure formed by the folding of multiple polypeptide chains
   d)   A structure where amino acids are linked by disulfide bonds

Q4    The interactions involved between distant amino acids within the protein sequence are called

      _____.
   a)   Primary structure
   b)   Secondary structure
   c)   Tertiary structure
   d)   Quaternary structure

Q5    The quaternary structure of a protein refers to _____.
   a)   The sequence of amino acids in the protein
   b)   The arrangement of alpha helices and beta sheets in the protein
   c)   The overall 3D shape of the protein
   d)   The arrangement of multiple protein subunits

Q6    _____ is the amino acid that is commonly involved in forming disulfide bonds in proteins.
   a)   Glycine
   b)   Alanine
   c)   Cysteine
   d)   Proline

Q7 Which of the following statement is true for the protein tertiary structure determination?
   a) The sequence of amino acids
   b) Hydrogen bonding between backbone atoms
   c) Ionic interactions between charged amino acids
   d) Overall folding and interactions of the secondary structures

Q8 Denaturation of a protein typically results in _____.
   a) Breaking of peptide bonds
   b) Loss of its 3D structure and function
   c) Formation of more stable hydrogen bonds
   d) Increased solubility in aqueous solutions

Q9 How X-ray crystallography is helpful in structural biology?
   a) By determining the sequence of amino acids in a protein
   b) By visualizing the 3D structure of proteins
   c) By analyzing protein–protein interactions
   d) By predicting the biological function of proteins

Q10 What does Bragg's law describe in the context of X-ray crystallography?
   a) The relationship between the wavelength of X-ray radiation and the diffraction angle
   b) The relationship between the atomic number of an element and its X-ray absorption
   c) The relationship between the size of a crystal and the intensity of X-ray diffraction
   d) The relationship between the temperature of a crystal and its X-ray diffraction pattern

**! Answers**
Q1 a) The sequence of amino acids
Q2 b) Hydrogen bonds
Q3 b) A coiled structure stabilized by hydrogen bonds between amino acids
Q4 c) Tertiary structure
Q5 d) The arrangement of multiple protein subunits
Q6 c) Cysteine
Q7 d) Overall folding and interactions of the secondary structures
Q8 b) Loss of its 3D structure and function
Q9 b) By visualizing the 3D structure of proteins
Q10 a) The relationship between the wavelength of X-ray radiation and the diffraction angle

# Index

https://doi.org/10.1515/9783111568584-010

www.ingramcontent.com/pod-product-compliance
Lightning Source LLC
Chambersburg PA
CBHW081529220326
41598CB00036B/6376